G000141427

Geophysical Monograph Series

Including

IUGG Volumes
Maurice Ewing Volumes
Mineral Physics Volumes

GEOPHYSICAL MONOGRAPH SERIES

IUGG Volumes

Maurice Ewing Volumes

Mineral Physics Volumes

Geophysical Monograph 47
IUGG Volume 2

Hydrogeological Regimes and Their Subsurface Thermal Effects

Alan E. Beck
Grant Garven
Lajos Stegena

Editors

American Geophysical Union

International Union of Geodesy and Geophysics

Geophysical Monograph/IUGG Series

Library of Congress Cataloging-in-Publication Data

Hydrogeological regimes and their subsurface thermal effects.

  (Geophysical monograph ; 47 /IUGG series ; 2)
  "The International Union of Geodesy and Geophysics Symposium U.S.
'Hydrogeological Regimes and Their Subsurface Thermal Effects' "—Foreword.
  1. Terrestrial heat flow—Congresses.   2. Groundwater flow—Congresses. I. Beck,
A. E.   II. Garven, Grant.   III. Stegena, Lajos,   IV. International Union of Geodesy
and Geophysics Symposium U.S. "Hydrogeological Regimes and Their Subsurface
Thermal Effects" (1987 : Vancouver, B.C.)   V. International Union of Geodesy and
Geophysics.   VI. American Geophysical Union.   VII. Series.

QE509.H93   1987   551.1'4   88-34792
ISBN 0-87590-451-3

Copyright 1989 by the American Geophysical Union, 2000 Florida Avenue,
NW, Washington, DC 20009

Figures, tables, and short excerpts may be reprinted in scientific books and
journals if the source is properly cited.

  Authorization to photocopy items for internal or personal use, or the
internal or personal use of specific clients, is granted by the American
Geophysical Union for libraries and other users registered with the Copyright
Clearance Center (CCC) Transactional Reporting Service, provided that the
base fee of $1.00 per copy plus $0.10 per page is paid directly to CCC, 21
Congress Street, Salem, MA 10970. 0065-8448/89/$01. + .10.
  This consent does not extend to other kinds of copying, such as copying for
creating new collective works or for resale. The reproduction of multiple
copies and the use of full articles or the use of extracts, including figures and
tables, for commercial purposes requires permission from AGU.

Printed in the United States of America.

CONTENTS

There are a number of reasons for organizing a symposium. One is to bring specialists together to discuss a specific problem in which they all have expertise and encourage the participants to talk openly about their current work which, in all likelihood, is not going to be in publishable form for some time. Another reason might be that a problem has been identified by one group of experts which needs the input from a previously loosely allied group before real progress can be made. The International Union of Geodesy and Geophysics Symposium U.8 "Hydrogeological Regimes and Their Subsurface Thermal Effects", organised for the August 1987 General Assembly in Vancouver, was a symposium of the latter type.

For many years, workers interested in the measurement and interpretation of terrestrial heat flow, although well aware that there were other and more efficient modes of heat transport than conduction, assumed that if temperature depth plots or total thermal resistance–depth (Bullard) plots were linear then perturbing influences were insignificant; in other words, linearity was a criterion of acceptability. Over the last few years it has become apparent that this criterion is simply not good enough in that one of the major uncertainties in interpretation of geothermal data is the extent of the influence of fluid movement. Few geothermal workers had much experience, or even knowledge, of problems faced by hydrologists and of the difficulties of inferring hydrogeological structure from field data; nor, until recently, were many hydrologists aware that the thermal effects may be used to constrain the hydrogeological structure. This symposium was conceived at the 1983 IUGG meetings in Hamburg by the International Heat Flow Commission, a commission of the International Association of Seismology and Physics of the Earth's Interior, during discussions of how to tackle this problem.

The principal objective of the symposium was, therefore, to initiate a closer contact between those members of the hydrogeological, hydrological and geothermal communities who wanted a better understanding of the sensitively coupled fluid-thermal system. It was not expected that major advances would be made at the symposium; rather, it was expected to be the catalyst that might lead to later significant advances in our understanding of the hydrogeological regimes and their thermal effects. It could, of course, lead to ways of making use of the thermal phenomena to map fluid flow fields. On the experimental side, we have to know what proportions of heat are being transported by phonon conduction and by fluid flow; understanding this has implications for the study of mantle heat flow patterns, basin development and the genesis of ore and hydrocarbon deposits. In the geothermal work to date we have only just begun to scratch the surface; for example, some interpretations of heat flow data require an assumption of fluid velocity fields and hence of permeability values. If one could confirm independently the inferred permeabilities for one or two regions, this might be turned into a technique for obtaining values of regional permeability elsewhere. It will be seen from some of the papers in this monograph that there are considerable differences of opinion about the direct and indirect approaches to determining permeabilities. On the other hand, many of the papers concerned with hydrology stress the flow paths and flow rates and tend to ignore the thermal effects. On the theoretical side, most work to date has assumed that a permeable layer is homogeneous and isotropic; clearly a move has to be made towards taking into account the spatial variability of permeability and to move from 2-D to 3-D modelling. Again, from some of the papers in this monograph it can be seen that a substantial start has been made but a great deal more has to be done.

In putting together the symposium, it was stressed that the speakers would not be pressed for a written publication for a proceedings unless most of them asked for it. In the event most, but not all, did prefer publication and this caused a small problem for the editors. A monograph should present as complete a picture as possible of the work going on in the subject area. Since one of the intents was to stimulate closer international cooperation between the geothermal and hydrological communities, there should be some papers dealing with an overview of various aspects of the subject, there should be some papers dealing with site or area specific case histories, and there should be some work reported that was as up to date as possible. The problem was that a number of speakers willing to give papers at the symposium were not willing to publish in the monograph – for a variety of understandable reasons, they had published the material already or something was already in the works, or the work was so up to date they could not prepare something in time for the deadlines. We have attempted to solve the first two parts of the problem by inviting extended abstracts, the main purpose of which is to summarize the work without detailed justification but with appropriate citations to which readers could turn if such justification was sought; for the last part there was nothing to do but leave gaps.

The editors wish to recognize the considerable help given them by the reviewers: S. Bachu, D. S. Chapman, M. J. Drury, J. P. Greenhouse, V. M. Hamza, A.M. Jessop, T.J. Lewis, L. Mansinha, C. W. Mase, P. Morgan, H. N. Pollack, J. H. Sass, P. Y. Shen, J. L. Smith, J. Toth, K. Wang, S. D. Willett, A. Woodbury.

Nearly all of them gave remarkably thorough reviews of the manuscripts and their constructive comments undoubtedly improved significantly the quality of the published manuscripts seen here.

The symposium on which this monograph is based would not have been possible without generous financial assistance from the Natural Sciences and Engineering Research Council of Canada with additional sums from the International Association of Seismology and Physics of the Earth's Interior, the International Union of Geodesy and Geophysics, and the International Association of Physical Sciences of the Oceans. It is a measure of the farsightedness of these organizations that they recognize that advances can be made by talking as well as by doing.

Alan E. Beck
Grant Garven
Lajos Stegena

NUMERICAL SIMULATION TECHNIQUES FOR MODELING ADVECTIVELY-DISTURBED THERMAL REGIMES

Leslie Smith[1], Craig Forster[2], and Allan Woodbury[3]

Abstract. Numerical models of heat redistribution by groundwater flow are helpful in examining the nature of the disturbance to a conductive thermal regime. Two aspects of this topic are discussed. First, we describe a modeling approach appropriate when simulating groundwater flow patterns and temperature distributions in mountainous terrain. Second, we outline a method of inverse simulation that uses temperature measurements to augment hydraulic head and hydraulic conductivity measurements in constructing models of subsurface flow systems. This latter technique exploits the sensitivity of the thermal field to hydrogeologic parameters.

Thermal Regimes in Mountainous Terrain

Ongoing interest in characterizing the earth's thermal state, investigating geologic processes within the upper crust, and exploring for geothermal resources has led to the collection of heat flow data in mountainous regions. A subset of these data indicates that groundwater flow can cause a significant advective disturbance of conductive thermal regimes [Steele and Blackwell, 1982; Mase et al., 1983; Black et al., 1983; Reader and Fairbank, 1983]. Groundwater flow systems in mountainous terrain differ from those in low-relief terrain in two key respects: (1) for a given set of hydrogeologic conditions, a greater range in water table elevation and form is possible, and (2) high-relief terrain enhances vertical components of groundwater flow and promotes circulation to depths where significant heating can occur.

Previous simulations of advective heat transfer have in general assumed a known water table configuration, and neglected fluid flow and heat transfer in the unsaturated zone [eg. Hanaoka, 1980; Smith and Chapman, 1983]. While appropriate when studying the thermal effects of regional groundwater flow in low-relief terrain; in mountainous terrain, this approach can lead to a poor representation of the system if hydrologic conditions promote the development of an extensive unsaturated zone. In this case, it is appropriate to treat the water table as a free surface, with fluid velocities and heat redistribution depending upon constraints imposed by infiltration rates, surface topography, and rock permeability. The location of the water table is estimated as part of the solution.

In the free surface approach, the upper boundary of the model domain is the bedrock surface. The boundary condition for fluid flow is expressed as an available infiltration rate, while the thermal boundary condition is an elevation-dependent mean annual ground surface temperature. The available infiltration is the maximum rate of recharge possible at the bedrock surface. It represents the difference between precipitation and runoff, where runoff is a lumped term accounting for surface water flow, evapotranspiration, and any subsurface flow through thin surficial deposits lying on the bedrock surface. In the absence of detailed site data, the available infiltration rate is best thought of as a percentage of the mean annual precipitation rate. Solutions of the coupled differential equations for hydraulic head and temperature can be conveniently obtained using finite element techniques [Forster and Smith, 1988a]. Conventional free surface techniques must be modified to account for the affects of steep topography [Forster, 1987].

Using this modeling approach, it is possible to examine how climatic, hydrogeologic and thermal regimes interact in mountainous terrain. Forster and Smith [1988b] focus on factors controlling patterns and magnitudes of groundwater flow. The two examples that follow illustrate important features of the thermal regime. Figure 1 shows an example of a conductive thermal regime in a system with 2 km of vertical relief over a lateral distance of 6 km. The convex topographic profile is representative of glaciated crystalline terrain. The bulk permeability of the mountain massif (k_u) is 10^{-18} m². A basal unit of reduced permeability (10^{-22} m²) is included within the solution domain. The basal heat flux is assumed to be 60 mW m⁻². Thermal conductivity of the rock mass is 2.5 W m⁻¹K⁻¹. For a given available infiltration of 2 x 10^{-9} m s⁻¹ (6 cm yr⁻¹), the water table occurs at the bedrock surface. Water in excess of that required to saturate the system is presumed to contribute to surface runoff. The heat lines show the transfer of the basal heat flux to the ground surface. These heat lines are the sum of the conductive and advective components of heat transfer. In this case, the conductive term is dominant.

[1]Department of Geological Sciences, The University of British Columbia, Vancouver, British Columbia

[2]Department of Geology, Utah State University, Logan, Utah

[3]Department of Geological Sciences, McGill University, Montreal, Quebec

Copyright 1989 by
International Union of Geodesy and Geophysics
and American Geophysical Union.

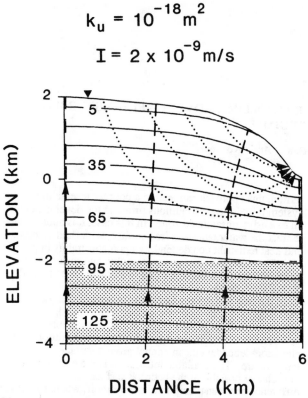

$$k_u = 10^{-18}m^2$$
$$I = 2 \times 10^{-9}m/s$$

Fig. 1. Example of a conductive thermal regime, showing isotherms in °C (solid lines) and heat lines (dashed lines). Dotted lines indicate pathlines for fluid flow. I is the available infiltration rate, k_u is permeability of the mountain massif above the basal low-permeability unit (stippled region). The inverted triangle identifies the water table.

Figure 2 shows the same mountain system, but for higher values of permeability. The available infiltration rate is unchanged from the previous case. Isotherms shown in the upper plot indicate a minor advective disturbance. Three features are shown on the lower plot, where the bulk permeability is increased to 10^{-15} m^2: (1) the system is permeable enough to transmit all the available infiltration, and as a result, the water table lies below the bedrock surface across much of the recharge area, (2) almost the entire basal heat flux is transferred to the valley by the groundwater flow system, and (3) the shaded region within the core of the mountain massif shows that for much of the region above the elevation of the valley floor, temperatures are governed by thermal conditions at the bedrock surface, and not by heat flow from below. Temperatures 2 km directly below the valley floor range from 90°C to 40°C as the bulk permeability of the mountain massif is increased from less than 10^{-18} m^2 to 10^{-15} m^2. A comparison of the disturbed regime shown in the lower plot of Figure 2 with the conductive regime of Figure 1 indicates that active groundwater flow cools almost the entire system. A region of warming is confined to a small area near the valley floor.

For sites where the water table lies below the bedrock surface, the threshold marking the transition from a conductive- to an advectively-disturbed regime can be

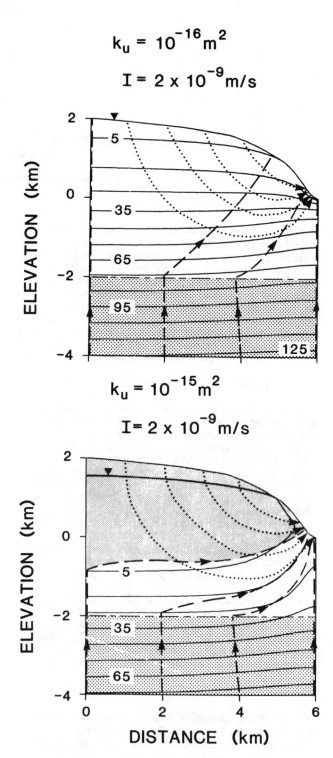

$$k_u = 10^{-16}m^2$$
$$I = 2 \times 10^{-9}m/s$$

$$k_u = 10^{-15}m^2$$
$$I = 2 \times 10^{-9}m/s$$

Fig. 2. The thermal regime for the system shown in Fig. 1, but with higher permeability values. The available infiltration rate I is fixed, k_u is permeability of the upper zone. Isotherms are in °C (solid lines). Dotted lines indicate pathlines for fluid flow. Heat lines (dashed lines) show the transfer of the basal heat flux to the ground surface. The inverted triangle identifies the water table.

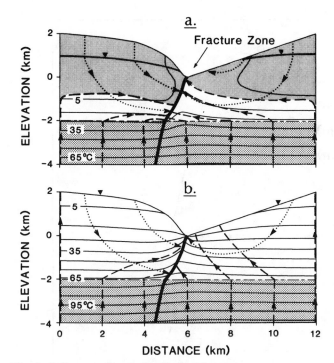

Fig. 3. Influence of a steeply dipping fracture zone on patterns of fluid flow and heat transfer in an asymmetric valley. Available infiltration rate is 2 x 10^{-9} m s^{-1}. k_u defines the permeability of the mountain massif above the basal, low-permeability unit; $k_f b$ is the product of fracture zone permeability and width of the fracture zone. (a) k_u=10^{-15} m^2, (b) k_u=10^{-16} m^2. Isotherms are in °C (solid lines). Dotted lines indicate pathlines for fluid flow. Heat lines (dashed lines) show the transfer of the basal heat flux to the fracture zone.

defined in terms of the available infiltration rate. Conductive thermal regimes are expected when available infiltration rates are less than about 10^{-11} m s^{-1}. The water table is most likely to lie below the bedrock surface in regions with higher permeability rocks (greater than 10^{-15} m^2) and arid to semi-arid climate. Where the water table coincides with the bedrock surface, the available infiltration rate no longer characterizes fluid flux within the rock mass. In such cases, an advective threshold can be defined on the basis of a bulk permeability value that is specific to the topographic relief of the domain.

Major fracture zones play an important role in the hydrology of mountain systems. Figure 3 illustrates the thermal effects of a fracture zone that daylights in the floor of an asymmetric valley. Two systems are shown; one with a bulk permeability (k_u) of 10^{-15} m^2, and the other with a value of 10^{-16} m^2. The more permeable system shows a strong advective disturbance. The fracture zone, with a width of 1 m, has a permeability 10^4 times greater than that of the mountain massif. The groundwater flow system directs the basal heat flux almost totally to the fracture zone. The volumetric flow rate in the fracture zone increases by approximately one order of magnitude along its length. The pattern of upwarped isotherms shown in Figure 3b differs

from those of Kilty et al. [1979] and Sorey [1978], who predict large temperature gradients at shallow depths in the fracture zone and almost isothermal conditions at greater depth because they specify uniform fluid within the fracture zone. Reduced fluid flux at depth in the fracture zones shown here causes reduced advective heat transfer at depth and produces a more uniform temperature distribution throughout the fracture zone.

Factors determining the discharge temperature of the thermal spring can be identified on Figure 4. Two curves are shown, one for a basal heat flux (H_b) of 60 mW m^{-2}, and the other 120 mW m^{-2}. There is a permeability "window" for which spring temperatures reach their highest values. As the value of the upper zone permeability is increased above the advective threshold, the spring temperature rises to a maximum at a permeability value of approximately 10^{-16} m^2. The higher spring temperatures occur as the fracture zone captures a greater proportion of the basal heat flux. However, for greater permeability values, spring temperature is observed to decline. Here the overall cooling of the mountain massif comes into play, as greater volumetric fluxes of groundwater absorb the thermal energy originating at the base of the domain with little increase in temperature.

For a given bulk permeability of the mountain massif, the transmissivity of the fracture zone ($k_f b$) determines the degree of the thermal disturbance. For the system shown in Figure 3, the fracture zone exerts its maximum influence when $k_f b$ equals about 10^4 k_u. Increasing $k_f b$ to 10^5 k_u has only a small incremental effect on the regional flow system, and causes only a 1°C increase in spring temperature (Figure 4). Reducing $k_f b$ to 10^3 k_u has a greater influence, leading to a 7°C decrease in spring temperature. Additional simulations show that the maximum spring temperature varies as the geometry of the fracture zone is modified. Interpretations of the temperature (and chemistry) of thermal springs that are not considered in a quantitative framework imposed by the regional hydrogeologic setting must be viewed with caution. Simple equations relating the geothermal gradient, depth of maximum circulation, and discharge temperature of thermal springs can potentially result in misleading conclusions.

Simultaneous Inversion of Hydrogeologic and Thermal Data

The examples just presented show that in more permeable systems, the thermal regime will reflect the influence of groundwater flow patterns and flow rates. Thus, temperature measurements are potentially useful in augmenting hydraulic head and hydraulic conductivity measurements when constructing models of hydrogeologic systems. To exploit this relationship, a simultaneous inversion technique is described that has as its goal the determination of medium parameters and uncertain boundary conditions, for both the fluid flow and thermal fields. For systems like those discussed in the previous section, uncertainties associated with estimates of bulk permeability can be as large as two or three orders of magnitude. A joint inversion of hydrogeologic and thermal data potentially leads to improved resolution of model parameters.

The approach is implemented by defining an objective function of the form:

$$J=(h-h^*)^t V_h^{-1}(h-h^*) + \lambda(Y-Y^*)^t V_Y^{-1}(Y-Y^*) + \beta(T-T^*)^t V_T^{-1}(T-T^*)$$

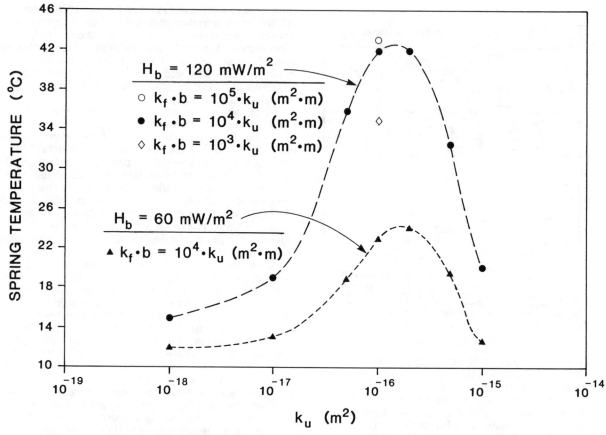

Fig. 4. Discharge temperature of the thermal spring for the fracture zone shown in Figure 3. Spring temperature is plotted as a function of the permeability k_u of the mountain massif, for two different values of basal heat flow (H_b). Also shown for the case with elevated heat flow are results for different transmissivities of the fracture zone ($k_f b$).

where h*, Y*, and T* are measured or interpolated values of hydraulic head, the logarithm of hydraulic conductivity, and temperature, respectively; h and T are corresponding nodal values of hydraulic head and temperature predicted using the model; V_h, V_Y, and V_T are prior hydraulic head, log conductivity and temperature covariance matrices, Y is a vector of log conductivity values to be determined from the inverse simulation, and λ and β are scaling factors. The prior covariance matrices describe the noise in the measurements. If the vectors h*, Y*, or T* are augmented by a spatial interpolation scheme such as kriging, then the prior covariance matrices also contain values of the estimation error. We seek the model that minimizes the difference between observed and predicted values of hydraulic head and temperature, conditioned on the available hydraulic conductivity data. More general forms of the objective function are possible when considering uncertainties in basal heat flux, thermal conductivity, or hydraulic boundary conditions. The objective function can be minimized using a constrained simplex technique [Woodbury, 1987].

Woodbury and Smith [1988] present a number of synthetic examples to evaluate the effectiveness of this joint inversion scheme. Steady state fluid flow and heat transfer is assumed. Improvements in parameter resolution depend upon the extent of the advective disturbance, which in turn is controlled by the magnitude of permeability values, the length scale and location of permeability variations, the depth of active flow, and the configuration of the groundwater flow system. Determination of parameters using this technique is enhanced for flow systems where the regional setting promotes vertical components of fluid flow. Because the magnitude of an advective disturbance depends upon absolute values of permeability, addition of temperature measurements in an inverse simulation is particularly helpful in cases where there is uncertainty in boundary conditions, such as recharge rates.

A field trial of this inverse technique is described in Woodbury and Smith [1988]. A cross-sectional model is constructed of groundwater flow within and beneath a landslide at a site located in south-central British Columbia. Forward simulations suggested the probable occurrence of advective heat redistribution. Hydraulic head and temperature data were collected from a series of multilevel piezometers installed across the landslide. A total of 31 measurements of hydraulic head and 96 temperature values formed the data base. Because prior hydraulic conductivity estimates are of doubtful reliability, these data are not included in the inverse simulation. There is also considerable uncertainty in some of the hydraulic head data because of long completion intervals in some piezometers. An inverse simulation of a homogeneous hydrogeologic system without thermal data could only lead to unique definition of the ratio between recharge rates and hydraulic conductivity. By including thermal data, bulk hydraulic conductivity values can be identified. Other parameters estimated include the depth of active flow, basal heat flux and thermal conductivity. Unfortunately, the full potential of this joint inversion approach is difficult to assess with this field trial because of the relatively poor quality of the hydrogeologic data. Additional investigations are planned at other sites to further evaluate the technique.

Acknowledgements. The work summarized in this extended abstract has been funded by grants from the Natural Sciences and Engineering Research Council of Canada.

References

Black, G.L., D.D. Blackwell, and J.L. Steele, Heat flow in the Oregon Cascades, in Geology and Geothermal Resources of the Central Oregon Cascades Range, ed. by G.R. Priest and B.F. Vogt, Oregon Dept. of Geol. and Min. Ind., Special Paper 15, 69-76, 1983.

Forster, C.B., Interaction of Groundwater Flow Systems and Thermal Regimes in Mountainous Terrain: A Numerical Study, Ph.D. Thesis, The University of British Columbia, Vancouver, Canada, 1987.

Forster, C.B., and L. Smith, Groundwater flow systems in mountainous terrain 1. Numerical modeling techniques, Water Resour. Res., 24(7), 999-1010, 1988a.

Forster, C.B., and L. Smith, Groundwater flow systems in mountainous terrain 2. Controlling factors, Water Resour. Res., 24(7), 1011-1023, 1988b.

Hanaoka, N., Numerical model experiment of hydrothermal system - topographic effects - Bulletin of the Geological Survey of Japan, 31(7), 321-332, 1980.

Kilty, K., Chapman, D.S., and C.W. Mase, Forced convective heat transfer in the Monroe Hot Springs geothermal area, Jour. Volc. and Geothermal Res., 6, 257-272, 1979.

Reader, J.F., and Fairbank, B.D., Heat flow in the vicinity of the Meager Volcanic Complex, Southwestern British Columbia, Geoth. Resour. Council Trans., Vol. 7, 535-539, 1983.

Smith, L., and D.S. Chapman, On the thermal effects of groundwater flow, Jour. Geophys. Res., 88(1), 593-608, 1983.

Woodbury, A.D., Simultaneous Inversion of Thermal and Hydrogeologic Data, Ph.D. Thesis, The University of British Columbia, Vancouver, Canada, 1987.

Sorey, M.L., Numerical modeling of liquid geothermal systems, U.S. Geol. Surv. Prof. Paper 1044-D, 1978.

Steele, J.L., and D.D. Blackwell, Heat flow in the vicinity of the Mount Hood Volcano, Oregon, in Geology and Geothermal Resources of the Mount Hood Area, Oregon, Oregon Dept. of Geol. and Min. Ind., Special Paper 14, 31-42, 1982.

Woodbury, A.D., and L. Smith, Simultaneous inversion of hydrogeologic and thermal data 2. Incorporation of thermal data, Water Resour. Res., 24(3), 356-372, 1988.

A SOLUTION TO THE INVERSE PROBLEM OF COUPLED HYDROLOGICAL AND THERMAL REGIMES

Kelin Wang, Po-Yu Shen and Alan E. Beck

Department of Geophysics, University of Western Ontario London, Canada, N6A 5B7

Abstract. In typical geological settings, the subsurface hydrological and thermal regimes are often closely coupled. A realistic analysis of the coupled systems requires that the two regimes be considered simultaneously. To make optimal use of the often noisy hydrological and thermal data, it is necessary to adopt an inverse formulation. In this paper, we report some results of our first stage investigation, using a steady state, 2–D (cross–section) model. A 2–D isoparametric finite element model is used to discretize the problem, and the nodal values of temperature and hydraulic head, as well as the elemental medium thermal conductivities and permeabilities, are treated as parameters. A generalized non–linear stochastic inverse method of Bayesian type is used for parameter estimation, with the a priori information on the parameters described in terms of the first two moments of the appropiate probability distributions. For computational efficiency, a gradient method is used in the parameter estimation procedure, and the gradient matrix (derivatives of the parameterized system with respect to the parameters), needed in the iteration scheme, is formulated analytically at the elemental level. Numerical results show that the non–linearity of the problem, which is effectively determined by the quality of the a priori information, plays an important role in the performance of the method. With a sufficient number of reasonably well distributed data, the parameters can be well resolved.

Introduction

Groundwater movement often provides a much more efficient means of subsurface heat transfer than pure heat conduction; the hydrological and thermal regimes have therefore been studied as coupled systems on various scales for different purposes. Groundwater is often used as part of a cooling system to "dump" waste heat, but thermal pollution could be a potential hazard to the environment and ecology; various aspects of this have been investigated by, e.g., Andrews and Anderson (1978, 1979), Sauty et al. (1982a, b), using numerical modeling techniques. In the petroleum industry, heat transport by

Copyright 1989 by
International Union of Geodesy and Geophysics
and American Geophysical Union.

flowing water is believed to have played an important role in the maturation and migration of oil (e.g., Hitchon, 1984; Gosnold and Fischer, 1986; Jones et al., 1985) and numerical modeling techniques have again been used in various studies (e.g., Garven and Freeze, 1984a,b; Garven, 1985, 1986; Doligez et al., 1986; Burrus and Bessis, 1986). In the study of terrestrial heat flow density, groundwater movement may perturb or even completely overshadow the desired component of conductive heat transfer. Regional scale heat transport by groundwater movement has been observed directly, or inferred in, for example, the Liaohe basin (Wang et al., 1985), the Uinta basin (Chapman et al., 1984), the Bohemian Cretaceous basin (Cermak and Jetel, 1985), and the Western Canada sedimentary basin (Jones et al., 1985); however, in at least one case, the inferred flow regime has been challenged on the basis that permeability values found from laboratory and field measurements are too low (Bachu, 1985). Smith and Chapman (1983) used a forward numerical model to investigate the problem of basin scale coupled thermal and hydrological regimes. Woodbury and Smith (1985), extended the model to 3–D. Because precise knowledge about the parameters and data is usually lacking in numerical modeling, inverse methods have been gaining attention in this field, and pioneer work has been done by Kasameyer et al. (1985) with an analytical model and Woodbury and Smith (1988) with a numerical model.

In groundwater hydrology, inverse groundwater modeling is a popular way to study isothermal flow problems and is synonymous with system identification. Aquifer parameters such as hydraulic conductivity, storativity, transmissivity, etc., are to be determined from hydraulic head data. There is an increasing number of publications in this field. Examples are: Cooley (1977, 1979), Neuman and Yakowitz (1979), Neuman et al. (1980), Yeh and Yoon (1981), Kitanidis and Vomvoris (1983), Yeh et al. (1983), Hoeksema and Kitanidis (1984), Dagan (1985), Sun and Yeh (1985), Carrera and Neuman (1986a, b, c), Fradkin and Dokter (1987), Loaiciga and Marino (1987).

In this work, we describe a method to obtain an inverse solution to the problem of basin wide coupled hydrological and thermal regimes. The problem, studied in 2–D, is first parameterized with an isoparametric finite element model, then a Bayesian type stochastic inversion is performed to give an optimal solution with statistical measures. In this preliminary work, synthetic examples using *a priori* information of varying quality are presented. Since the purpose of the paper is to provide a description of the method for readers with general interests, we introduce as little mathematical detail as possible, leaving the detailed mathematical development of the method for presentation elsewhere (Wang and Beck, 1989).

We pose the inverse problem in a slightly different way from that of system identification. We often have data or some kind of knowledge about all the involved physical quantities, be they material properties or field variables, dependent or independent, i.e, we know something about everything. Our task is to use the available information and relevant physical laws to obtain a better idea of the whole picture. For this reason, "inverse" here has a more general meaning than inverting field variable data to determine material properties; the method is really an optimal finite element solution to heat transfer problems in a heterogeneous porous medium.

Our work has some similarities to that of Woodbury and Smith (1988) who used thermal data as additional constraints to the problem of hydrological system identification. Both treatments deal with hydrologic and thermal regimes simultaneously, use a 2–D finite element model to discretize the problem, and solve a nonlinear inverse problem. The major differences are: 1) we consider a larger depth scale, and hence a wider temperature range, so that the temperature dependence of water density and dynamic viscosity must be taken into account, therefore the two partial differential equations describing water flow and heat transfer are nonlinear as well as coupled; 2) in this work, the inverse method is more general in the sense that not only the material properties but also the field variables are treated as parameters and 3) at the parameter estimation stage, we use a gradient method.

Parameterization

Mathematical Model

Our problem is confined to a low Reynolds number, slightly compressible, two dimensional Darcian flow of sedimentary basin scale. For non–isothermal subsurface flow problems, it is convenient and customary to define a reference hydraulic head

$$h = \frac{P}{\rho_0 g} + x_2 \qquad (1)$$

where P is water pressure, ρ_0 water density at a reference temperature T_0 and g the absolute value of gravitational acceleration, with x_2 axis pointing upwards.

Then, in a heterogeneous, isotropic porous medium, Darcy's law takes the form,

$$v_i = -\frac{\kappa}{\mu}\rho_0 g \left(\frac{\partial h}{\partial x_i} + \rho_r \delta_{2i}\right) \qquad (2)$$

where v_i is specific discharge or Darcian velocity of water, κ the permeability of the medium, a function of x_i, μ dynamic viscosity of water, δ_{ij} Kronecker delta, and ρ_r the relative water density, defined as

$$\rho_r = \frac{\rho - \rho_0}{\rho_0} \qquad (3)$$

ρ being the water density at temperature T.

For a slightly compressible steady state Darcian flow, the following continuity equation is sufficiently accurate (e.g., Bear, 1972),

$$\frac{\partial}{\partial x_i}\left\{\frac{\kappa}{\mu}\left(\frac{\partial h}{\partial x_i} + \rho_r \delta_{2i}\right)\right\} = 0 \qquad (4)$$

Here, and in the rest of the paper, repeated subscripts imply summation unless specified otherwise.

For the same heterogeneous isotropic medium, Fourier's law of heat conduction states that,

$$q_i = -\lambda \frac{\partial T}{\partial x_i} \qquad (5)$$

where q_i is heat flow density, T is temperature, and λ is the overall (bulk) thermal conductivity of the fluid–solid complex.

In the presence of flowing pore fluid, there is also a dispersive component in the convective heat transfer process, but it is usually considered insignificant (Bear, 1972, p.651; Mercer et al., 1975; Woodbury and Smith, 1985). We simply lump all effects into one coefficient as suggested by Mercer and Faust (1980), and call it the thermal conductivity, but keeping in mind that it includes the effect of thermal dispersion, if the latter is not negligible.

The steady state energy balance equation in our case is then,

$$\frac{\partial}{\partial x_i}\lambda \frac{\partial T}{\partial x_i} - \rho c v_i \frac{\partial T}{\partial x_i} + Q = 0 \qquad (6)$$

where Q is a source term, such as heat produced by radioactive elements, and c is the specific heat capacity of water.

Equations (4) and (6), subject to appropriate boundary conditions, constitute the mathematical model of our problem. They are coupled and the coupling is due to the temperature dependence of ρ and μ in (4) and the convective term in (6).

The temperature (and hence depth) dependence of μ has a direct influence on the magnitude of the Darcian velocity, which in turn determines the relative importance of convective heat transport with respect to the conductive component. Over the temperature range relevant to our problem (20 °C ~ 150 °C), μ varies by a factor of 4 ~ 5 and must be treated as a function of temperature. The temperature dependence of ρ_r gives rise to a buoyancy term in (4), which may have a similar magnitude to that of the head gradient, and must be taken into account. However, the variation of the specific thermal capacity ρc, in (6), is only a few percent over the same range and is negligible compared to that of μ.

A number of precise expressions for ρ and μ as functions of T exist in the literature (e.g., Mercer et al., 1975; Straus and Schubert, 1977), but the following linear approximations illustrated in Fig.1, which have the merit of being convenient for analytical manipulation, are adopted in this work:

$$\rho = \rho_0 - \beta(T - T_0) \qquad (7)$$
$$\mu^{-1} = \mu_0^{-1} + \eta(T - T_0) \qquad (8)$$

where β and η are constants, the value of which are given in Table 1 (together with the values of other parameter constants used). It is easy to see that if better accuracy for ρ and μ or wider temperature range is required when using our inverse method, it is not difficult to replace (7) and (8) by multi–section linear forms.

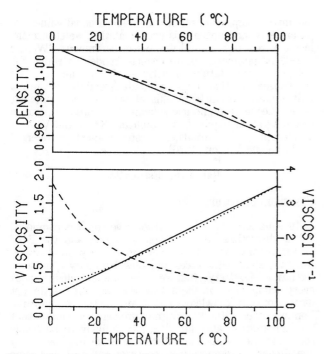

Fig.1. Density ρ (g cm^{-3}) and dynamic viscosity μ (kg s^{-1}m^{-1}) of water as functions of temperatures. Dashed lines represent ρ and μ as given by Bejan (1984, p.464); dotted line is μ^{-1}. The linear approximations of ρ and μ^{-1} are shown by solid lines.

Outline of Finite Element Formulation

In common with most inverse theories in earth sciences, our method involves parameterization and eventually becomes a problem of statistical parameter estimation. As pointed out by Tarantola and Valette (1982), it is often difficult to distinguish between data and the "parameters" in theoretical physical models; all the uncertain quantities have more or less the same positions in a general inverse problem. This point of view is of particular relevance to our inverse method; it is only of idiomatic convenience that we distinguish between the commonly measured field variables temperature and hydraulic head, and the more sparsely measured material properties thermal conductivity and permeability. The parameterization is performed with a finite element model, in which the material properties are discretized in a stepwise fashion and the field variables are represented by their nodal values.

The spatial domain is divided into a number of elements, each having N_e nodal points and a uniform value of λ and of κ. T and h in each element are interpolated using their point values T_k and h_k at the elemental nodes and polynomial interpolation functions $H(x)$, i.e.,

$$T = H_k T_k \qquad (9a)$$
$$h = H_k h_k \qquad (9b)$$

where the subscript k represents the global nodal number

but the summation is made over all the N_e elemental nodes.

For the coupled system (4) and (6), a forward finite element solution can be most efficiently obtained with a sequential solution procedure (Huyakorn and Pinder, 1983, p.198) which solves the two equations alternately, each time using the updated ρ and μ in (4) and v in (6) as functions of space, but not of temperature or head, thus making the equations linear, until both field variables, T and h, converge. Because the scheme has the nice property of using minimal computer memory and is easy to program, it was used exclusively in all the forward finite element numerical modeling referred to in the introduction. The scheme is also applicable to a non–linear inverse problem, if a gradient method is not used in parameter estimation or if the gradients are obtained numerically. In order to use a computationally efficient analytical gradient method, a more suitable finite element coupled solution procedure was chosen for this work.

The coupled solution procedure solves the two nonlinear partial differential equations (4) and (6) simultaneously. We define a vector U in the (T,h) plane,

$$U = \begin{bmatrix} T \\ h \end{bmatrix} \qquad (10)$$

Equation (9) is then written

$$U = \begin{bmatrix} H_k & 0 \\ 0 & H_k \end{bmatrix} \cdot U_k \qquad (11)$$

where $U_k = [\, T_k \;\; h_k \,]^t$; superscript t represents transpose operation. Each nodal point now has two degrees of freedom, T and h, which are rather like the two components of displacement or velocity in a 2–D mechanics problem.

Using the Galerkin weighted residual method (e.g., Zienkiewicz, 1972) and following a procedure similar to the derivation of the coupled solution procedure for the problem of single fluid heat transfer in a porous medium by Huyakorn and Pinder (1983, p.204), we obtain the following algebraic system,

$$K^e \cdot U = f^e \qquad (12)$$

where the "stiffness" matrix K is referred to as the conductivity matrix with the term conductivity implying both thermal conductivity and hydraulic conductivity and f is the equivalent nodal flow (ENF) vector. Superscript e implies that K^e and f^e are elemental, but they all have the global dimension. U, the global nodal temperature–head vector, has all the nodal values of T and h as its components.

TABLE 1. A list of parameter constants

Constant	Value	Unit
T_o	30	^oC
ρ_o	995.91	kg m^{-3}
β	0.53625	kg m^{-3} K^{-1}
μ_o^{-1}	1253.1	s m kg^{-1}
η	32.579	m s kg^{-1} K^{-1}
ρc	4.18 x 10^6	J m^{-3} K^{-1}
g	9.8	m s^{-2}

K^e is a function of T, h, λ and κ. f^e consists of two parts,

$$f^e = f_q^e + f_Q^e \qquad (13)$$

where f_q^e is the contribution from boundary heat and water flow inputs, and f_Q^e is the contribution from heat sources.

The boundary flow inputs are normal to elemental boundaries and are specified by their point values q_k at elemental boundary nodes. The point values form a nodal flow vector q, which includes both heat and water flux. Obviously, non-zero values can occur only at nodal points on the global boundary. f_q^e is related to q by a linear transformation,

$$f_q^e = V^e \cdot q \qquad (14)$$

where matrix V^e is determined by the elemental interpolation function $H(x)$.

Body heat source is taken as a constant in each element and is linearly transformed into f_Q^e in a similar way.

The global system

$$K \cdot U = f \qquad (15)$$

is the combination of the elemental systems (12) over all elements, with

$$K = \sum_e K^e \qquad (16)$$

$$f = \sum_e f^e \qquad (17)$$

Since K is a function of U as well as of λ and κ, equation (15) is a nonlinear static system. To meet the requirements of the subsequent parameter estimation procedure, in which λ and κ are assumed to be log-normally distributed, we introduce the following variables

$$\gamma = \ln\lambda \qquad (18)$$
$$\psi = \ln\kappa \qquad (19)$$

as the new parameters which are normally distributed. In the nonlinear system (15), U and the discretized γ and ψ are to be estimated. To reduce the total number of parameters, elements are grouped into a number of material property zones, so that a discretized value of λ or κ is usually the average over a group of elements (zonation). Arranging all the components of U and the discretized γ and ψ to form a parameter vector p, we rewrite (15) as

$$g(p) = f \qquad (20)$$

This is the final parameterized mathematical model of the inverse problem.

The above procedure is general for any finite element model. In an actual forward implementation, some authors favour lower degrees of interpolation combined with more nodes and elements, while others prefer higher degrees of interpolation combined with fewer nodes and elements. The extensively used linear triangular element model is an example of the former; it has the advantage that K^e, f^e and V^e can be readily obtained without performing numerical integration. However, large numbers of nodes and elements have to be used to maintain satisfactory accuracy of the discretization. In our inverse approach, all of the T and h nodal values are parameters, and hence it is expedient that we choose the latter to reduce the total number of parameters. We use a 2-D isoparametric finite element model, in which the set of interpolation functions H define a 1-1 transformation that maps a quadrilateral in $\mathbb{R}^2$ onto a unit square also in $\mathbb{R}^2$. This allows the most efficient Gauss-Legendre numerical scheme to be used to perform the integration required in computing K^e, f^e and V^e over all elements. For details, interested readers are referred to Bathe and Wilson (1976).

Parameter Estimation

Bayesian Estimation

In equation (20), the parameter vector p is composed of the discretized T, P, γ and ψ values, and the ENF vector f contains values that are linearly transformed from the specified sources and flow boundary conditions (equation 14). Very often we have some knowledge about every component of these two vectors, but we do not have complete knowledge about any of the components. For example, without any measurement, we know that the temperature at 1 km depth is unlikely to be below zero; on the other hand, even with a high accuracy instrument, a temperature measurement still has some uncertainties. In Bayesian statistical inference (Box and Tiao, 1973) this state of information is conveniently depicted by regarding p and f as random vectors obeying certain probability distributions. If the probability density function (PDF) of a random vector r is denoted by $\mathscr{P}(r)$ and the conditional PDF of r_1 given r_2 by $\mathscr{P}(r_1|r_2)$, Bayes' rule states that

$$\mathscr{P}(p|f) = \frac{\mathscr{P}(f|p)\,\mathscr{P}(p)}{\mathscr{P}(f)} \qquad (21)$$

$\mathscr{P}(p)$, which tells us what is known about p without knowledge of f, is called the *a priori* PDF of p; $\mathscr{P}(p|f)$, which tells us what is known about p given knowledge of f, is called the *a posteriori* PDF of p. We define $\mathscr{P}(p)$ in an arbitrary manner using the available information on p based on observation, previous study or even personal experience and engineering judgment. Here the word "arbitrary" is used to mean that physical laws such as eq.(20) are not necessarily involved. On the other hand, by definition of conditional probability, $\mathscr{P}(f|p)$, the PDF of f given p does involve the physical laws. If $\mathscr{P}(f|p)$ is regarded as a function of p, not of f, it represents the likelihood for p given f, and is thus called the likelihood function for p. If the error term

$$e_f = f - g(p) \qquad (22)$$

is assumed to have zero mean $E(e_f) = 0$, then clearly

$$\mathscr{P}(f|p) = \mathscr{P}(e_f) \qquad (23)$$

The denominator $\mathscr{P}(f)$ in (21) can be shown to be a constant and of no importance in the estimation of p (Box and Tiao, 1973). Bayes' rule (21) provides a mathematical formulation of how knowledge from different sources are combined to give new knowledge. The knowledge of f, in the form of $\mathscr{P}(f|p)$, consists not only of the values and the uncertainties of f, but most importantly, also of the physical laws governing the relation between **f** and **p**, and between different components of **p**.

In this work we make the assumption that $\mathscr{P}(\mathbf{p})$ and $\mathscr{P}(\mathbf{e_f})$ are Gaussian, with expectations $\mathbf{p_o}$ and $\mathbf{0}$, and covariance matrices C_{pp} and C_{ff}, respectively. The *a posteriori* PDF defined by (21) is then,

$$\text{const} \cdot \exp -\frac{1}{2}\{[\mathbf{f} - \mathbf{g}(\mathbf{p})]^t \, C_{ff}^{-1} \, [\mathbf{f} - \mathbf{g}(\mathbf{p})]$$

$$+ \, [\mathbf{p} - \mathbf{p_o}]^t \, C_{pp}^{-1} \, [\mathbf{p} - \mathbf{p_o}]\} \qquad (24)$$

Support for the Gaussian assumption includes evidence such as the well recognized log normal distribution of subsurface material permeability (Freeze, 1975) and the log normal–like shape of most of the thermal conductivity histograms compiled by Roy et al. (1981).

In Bayesian inference, only through a thorough study of the *a posteriori* PDF, can our updated knowledge of the parameters be sufficiently described (Box and Tiao, 1973). For a linear inverse problem based on Gaussian distribution, the *a posteriori* PDF is also Gaussian and can be completely characterized by its expectation and covariance matrix. In that case, it is justified to use a point estimate, namely the expectation, to represent the *a posteriori* PDF. For a nonlinear inverse problem, a point estimate is generally not representative of the usually complex behaviour of the non–Gaussian *a posteriori* PDF. However, for a large multivariate problem, the computation of the *a posteriori* PDF is usually an insurmontable task and the presentation of the results would also be difficult. For these reasons, to seek point estimates is still a common practice in Bayesian type nonlinear inversion (e.g., Gavalas et al., 1976; Jackson and Matsu'ura, 1985); in fact, the term "inverse method" in the most popular interpretation implies point estimates, but due to the complexity of the *a posteriori* PDF, the choice of a particular estimator is usually a matter of computational convenience, there being no general selection rule (Schweppe, 1973). Among the four types of Bayes estimator summarized by Schweppe (1973), the "most probable" type, which searches for the point $\hat{\mathbf{p}}$ in the parameter space that maximizes $\mathscr{P}(\mathbf{p}|\mathbf{f})$, is used in this work because it is the easiest to compute, and gives the most physically meaningful point estimate if there has to be one. The validity of the most probable estimate depends inversely upon the nonlinearity of the problem; for a linear problem, the most probable estimate coincides with the *a posteriori* expectation; for a slightly nonlinear problem, the most probable estimate can be used to approximate the *a posteriori* expectation; for a very nonlinear problem, however, the most probable estimate may bear no relation to the expectation and caution should be taken in interpreting the results of point estimation (e.g., Jackson and Matsu'ura, 1985).

Adopting the most probable estimate, our problem becomes one of optimization with the objective function

$$\Pi = [\mathbf{f} - \mathbf{g}(\mathbf{p})]^t \, C_{ff}^{-1} \, [\mathbf{f} - \mathbf{g}(\mathbf{p})]$$

$$+ \, [\mathbf{p} - \mathbf{p_o}]^t \, C_{pp}^{-1} \, [\mathbf{p} - \mathbf{p_o}] \qquad (25)$$

We seek the point $\hat{\mathbf{p}}$ at which Π reaches the global minimum.

Optimization Scheme

The objective function defined by (25) can be minimized using various gradient methods which use the partial derivatives of Π with respect to the components of **p**, or non–gradient methods which compute Π directly. The simplex technique, a classical non–gradient method (e.g., Beveridge and Schechter, 1965), was used by Woodbury and Smith (1988) with some modifications in a simultaneous finite element inversion of hydrogeologic and thermal data. In addition to not having to compute the gradients, an obvious advantage of the simplex technique is that it can be readily applied to objective functions constructed with different norms. For example, it is much easier to adapt the simplex technique to L_1–norm minimization, a criterion which is more robust than the L_2–norm minimization in the sense that a few bad data points (outliers) would not alter the solution drastically (Woodbury et al., 1987). A disadvantage of the approach is that numerous forward computations must be performed to solve equations (4) and (6). In the work of Woodbury and Smith (1988), ρ and μ are treated as constants because the temperature range considered is small. As a result, equations (4) and (6) are decoupled and linear, and need be solved only once for each forward calculation. For a coupled nonlinear problem, each forward computation requires, typically, 10 iterations for convergence, resulting in a 10–fold increase of computer time. For this reason, a gradient method which speeds convergence is preferred in this work.

We adopt the scheme derived by Rodgers (1976) and Tarantola and Valette (1982),

$$\hat{\mathbf{p}}_{k+1} = \mathbf{p_o} + C_{pp} \cdot G_k^t \cdot \mathbf{Y_k} \qquad (26a)$$

$$\mathbf{Y_k} = R_k^{-1} \cdot [\mathbf{f_o} - \mathbf{g}(\hat{\mathbf{p}}_k) + G_k \cdot (\hat{\mathbf{p}}_k - \mathbf{p_o})] \qquad (26b)$$

where bold faced subscript **k** is the iteration step number, and G_k is the gradient matrix G at the kth iteration,

$$G_k = (G_{lm})_k = \left(\frac{\partial g_l}{\partial p_m}\right)\Big|_{\mathbf{p} = \hat{\mathbf{p}}_k} \qquad (27)$$

and

$$R_k = C_{ff} + G_k \cdot C_{pp} \cdot G_k^t \qquad (28)$$

Obviously, the symmetric matrix R is positive definite, and a very efficient method can be used to solve the linear algebraic system (26b). As shown by Rodgers (1976), Tarantola (1987) and Wang and Beck (1987), the scheme is asymptotically a Newton's iteration algorithm. Newton's iteration converges quadratically to a unique solution only when the problem is mildly nonlinear and when the initial parameter value is reasonably close to the solution value (Burden et al., 1981). The nonlinearity and the closeness of the initial value to the solution value depend mostly on the quality of the *a priori* information on **p** and **f**.

In (isothermal) groundwater hydrology, the "gradients" are commonly the partial derivatives of head with respect to the aquifer transport parameters to be estimated, such as the transmissivities and storativities, and are called the sensitivity coefficients. The computation of sensitivity coefficients is usually a tedious task, especially for coupled problems such as the one in this study. A summary of three types of method commonly used for computing sensitivity coefficients was given by Yeh (1986). In this work, as has been stated earlier, we define parameters in a more general sense; parameter vector $\mathbf{p}$ includes $\mathbf{T}$ and $\mathbf{h}$ as well as discretized logarithmic material properties. Instead of working with the sensitivity coefficients, we work with the gradient matrix G, which can be derived analytically with basic calculus rules.

At first glance, it might appear that the function $\mathbf{g}$, defined by (20), is rather difficult to differentiate, but when we realize that

$$\mathbf{g}(\mathbf{p}) = KU = (\sum_e K^e)U = \sum_e (K^e U) = \sum_e \mathbf{g}^e(\mathbf{p})$$

$$(29)$$

we can derive G at the elemental level, i.e.,

$$G^e_{lm} = \frac{\partial g^e_l}{\partial p_m} \qquad (30)$$

then use

$$G = \sum_e G^e \qquad (31)$$

to assemble the global G matrix. In the inverse solution procedure both G and K matrices are formulated at the elemental stage and assembled at the global stage for each iteration.

Covariance Matrices

A priori covariance matrix of $\mathbf{p}$. If the *a priori* information is based on a previous statistical or stochastic analysis such as kriging (e.g., Delfiner, 1976; Delhomme, 1978) or on a previous inversion (e.g., with the method of Willett and Chapman, 1987), the structure of C_{PP} is given. Without any statistical study, C_{PP} is conveniently assumed to be diagonal, meaning that the components of $\mathbf{p}$ are considered mutually uncorrelated *a priori*. At first sight, this assumption may seem puzzling because, except for synthetic examples and for those cases where all the components of $\mathbf{p}$ are measured *a priori* with well calibrated equipment and where the measured quantities do not have much irregular spatial variations, the sub-diagonal entries of C_{PP} are generally non-zero; but it should be remembered that the *a priori* PDF of $\mathbf{p}$ is only a summary of the available information defined in an arbitrary manner to describe our knowledge or ignorance of the parameters. If all we know about the parameters is their approximate values with variances, then the covariance matrix C_{PP} has to be diagonal. If we do have knowledge about the covariances but still assume a diagonal C_{PP}, maybe for computational efficiency, there is a waste of information because we do not make full use of the knowledge we have. In any case, the shape of C_{PP} is not as critical a factor in our approach as the generalized covariance function in kriging or the correlation function in the method of Willett and Chapman. In kriging, the physical quantities are treated as spatial stochastic processes, and the generalized covariance function, which describes the stationarity features of the processes, is the only law governing the spatial relation between the values of quantities at different points, and hence the results depend solely on

the performance of the function; similar comments apply to the correlation function regarding the horizontal relation of temperatures in the method of Willett and Chapman. In our approach, physical laws such as Darcy's law and Fourier's law represented by (20) are explicitly introduced by the likelihood function to control the (deterministic) spatial relation of the physical quantities. A C_{PP} that accurately describes the correlation between the errors in the *a priori* parameter values is desirable, and lacking this information will introduce some bias in the solution; but the bias is minimized if there are other kinds of *a priori* information, since the *a priori* PDF of $\mathbf{p}$ will be "modified" by the physical laws.

A posteriori covariance matrix of $\mathbf{p}$. The *a posteriori* covariance matrix $\hat{C}_{PP}$ is defined as (Schweppe, 1973)

$$\hat{C}_{PP} = E[(\mathbf{p} - \hat{\mathbf{p}}) \cdot (\mathbf{p} - \hat{\mathbf{p}})^t] \qquad (32)$$

For a linear problem

$$\mathbf{f} = \mathbf{g}(\mathbf{p}) = G \cdot \mathbf{p} \qquad (33)$$

where G is a constant matrix, if the *a priori* PDF of $\mathbf{p}$ and of $\mathbf{e}_f$ are Gaussian, the *a posteriori* PDF $\mathscr{P}(\mathbf{p}|\mathbf{f})$ is also Gaussian, and the most probable estimate is identical to the conditional expectation. In that case, the *a posteriori* $\hat{C}_{PP}$ is given by (Schweppe, 1973; Rodgers, 1976; Tarantola and Valette, 1982):

$$C_{PP} = C_{PP} - C_{PP} \cdot G^t \cdot (C_{ff} + G \cdot C_{PP} \cdot G^t)^{-1} \cdot G \cdot C_{PP}$$

$$(34)$$

For the nonlinear system (20), the *a posteriori* PDF is generally not Gaussian and it would be very difficult to develop $\hat{C}_{PP}$. The usual practice is to linearize (24) in the neighbourhood of the most probable estimate point $\hat{\mathbf{p}}$, approximate the conditional expectation with $\hat{\mathbf{p}}$, and use (34) as a linear approximation of the *a posteriori* covariance matrix (Jackson and Matsu'ura, 1985; Vasseur et al., 1986). In fact, Rodgers (1976) derived iteration scheme (26) by linearizing $\mathbf{g}(\mathbf{p})$ in the first place.

The covriance matrix of $\mathbf{e}_f$. The likelihood function $\mathscr{P}(\mathbf{f}|\mathbf{p})$ is given by (23). The error vector $\mathbf{e}_f$ consists of three parts,

$$\mathbf{e}_f = \mathbf{e}_q + \mathbf{e}_Q + \mathbf{e}_m \qquad (35)$$

where $\mathbf{e}_q$ is the error due to uncertainties in the boundary conditions, $\mathbf{e}_Q$ is the error due to uncertainties in the heat sources, and $\mathbf{e}_m$ is a model error mainly due to the inaccuracy of the invoked physical laws and the finite element discretization of the problem.

$\mathbf{e}_Q$ may vanish when heat sources do not exist. The components of $\mathbf{e}_q$ vanish identically at inner finite element nodes and at any impermeable (to heat or/and water) boundary. $\mathbf{e}_m$ is usually very small but non-zero because the finite element discretization always has approximation error, though theoretically the error approaches zero when a very fine element mesh is used. In ideal cases $\mathbf{e}_m$ may vanish, e.g., when synthetic data, generated with a forward finite element solution with the same mesh as for the inverse solution, are used. Therefore, C_{ff} is generally positive semi-definite.

The covariance matrix of updated $\mathbf{f}$. We define the covariance matrix of updated $\mathbf{f}$ as

$$\hat{C}_{ff} = E[(f - \hat{f}) \cdot (f - \hat{f})^t] \qquad (36)$$

where $\hat{f}$ is the modified value of f at $\hat{p}$,

$$\hat{f} = g(\hat{p}) \qquad (37)$$

From (36), it is obvious that for a linear problem such as (33),

$$\hat{C}_{ff} = G \cdot \hat{C}_{pp} \cdot G^t \qquad (38)$$

For mildly nonlinear problems, (38) can be used as a linear approximation for $\hat{C}_{ff}$.

About Boundary Flows

$\hat{f}$ can be used to compute the boundary heat flux and water flux as modified by the inversion. This is especially desirable if we are interested in the background heat flow density values. It is generally difficult to separate the contributions from the boundary flux and the source term. Either q or Q must be well constrained for the other to be resolved. The theory for updating the boundary fluxes in the absence of heat sources is straight forward, as is discussed below; the computation of $\hat{q}$, the updated boundary flux vector, and its covariance matrix $\hat{C}_{qq}$ is presented in Wang and Beck (1989).

The contribution of nodal flow vector q to the ENF vector f is given by the linear transformation (14). If we reduce the dimensions of q, f and V by discarding all the components associated with the internal element nodal points and the impermeable (to heat or water) boundary nodal points, matrix V will be nonsingular and $\hat{q}$ and $\hat{C}_{qq}$ can be obtained as

$$q = V^{-1}\hat{f} \qquad (39)$$

and

$$\hat{C}_{qq} = V^{-1} \cdot \hat{C}_{ff} \cdot (V^{-1})^t$$
$$= V^{-1} \cdot G \cdot \hat{C}_{pp} \cdot G^t \cdot (V^{-1})^t \qquad (40)$$

Synthetic Examples

General Remarks

Any new method of parameter inference should first be tested against a well known parameter set. The best control of all comes from a synthetic model and we use this approach to illustrate some general features of the solutions. First, a geologic model is assumed with known permeability and thermal conductivity values, and specified boundary conditions, the temperature and head are then obtained by using a forward calculation. These λ, κ, T and h values compose the true parameter vector, which is then perturbed with simulated noise of known probability distribution, to form the a priori data set for the inverse problem. Only one synthetic geologic model is considered in this paper but by varying the noise level of the data set or of different parts of the data set, it is possible to study the reliability of the solutions in

relation to the quality of the a priori information; this in turn gives some idea as to how reliable the field data must be if interpretations with a specified confidence level are to be obtained. The synthetic examples are meant to illustrate the method, not to simulate a real geological model. A real model can be approached by increasing the number of material property zones.

The geologic model in this paper is a tectonically undisturbed sedimentary basin in the form of a trough which allows a 2–D representation in the vertical plane perpendicular to its strike. In cross–section (Fig.2) the model is composed of four geologic units with a layered structure, with the rock material of each layer characterized by a distinct permeability value. Also, the two upper layers and the two lower layers form two thermal conductivity zones, each having a uniform λ value. The material property values for the permeability and conductivity zones are listed in Table 2. No heat sources exist.

At the four boundaries, we specify the following conditions for the forward solution. The two vertical boundaries of the cross–section are assumed to have the "symmetry" condition, i.e., there is neither heat flow nor water flow across the boundaries. The upper boundary is the water table, i.e., the fresh water hydraulic head at a point on the boundary takes the same value as the elevation (free surface); this boundary is also taken as the 30 °C isotherm. The lower boundary separates the permeable rocks and the underlying impermeable basement and across which there is a vertical constant heat flow density of 60 mW m^{-2} from below. These boundary conditions, similar to those used and discussed by Smith and Chapman (1983), are considered reasonable for basin scale problems.

The forward calculation was performed with a 2–D isoparametric finite element method using the sequential solution procedure (Huyakorn and Pinder, 1983, p.198). The cross–section was divided into 32 quadrilateral elements, each having eight nodes, giving rise to a total of 121 nodal points; the finite element layers were chosen to coincide with the permeability zones. The same element mesh has been used for both the forward and inverse solutions, therefore the nodal values of T and h given by the forward solution can be used directly as the true values of the global temperature–head vector U in (13). The results of the forward solution are shown in Fig.2b.

The boundary conditions for the inverse problem are not exactly the same as those of the forward problem. The zero flow inputs across the two vertical boundaries are specified without uncertainty, as are the zero water flow input from the underlying basement. At the lower boundary, the background heat flow density of 60 W m^{-2} is taken as well constrained with a small error bar of 2%. The top boundary, however, has specified field variable value (Dirichlet) conditions for both T and h in the forward calculation, and the boundary fluxes are computed as part of the forward solution. With this synthetic model, we apply moderately uncertain heat flow density and water discharge rates on this boundary for the inversion. The output flux values of the forward solution are more conveniently given at the numerical integration sample points of each element than at the

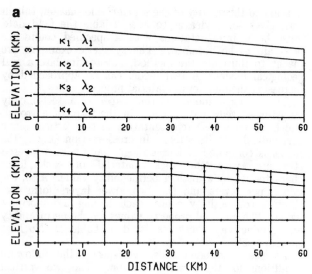

Fig.2a. A simple synthetic geologic model and the finite element mesh used for both the forward and inverse solutions. The medium is heterogeneous and consists of four layered geologic units, each of which has a distinct permeability value, but the top two units and the bottom two units form two thermal conductivity zones. The conductivity and permeability values are listed in Table 2. The finite element discretization in the vertical direction coincides with the geologic units.

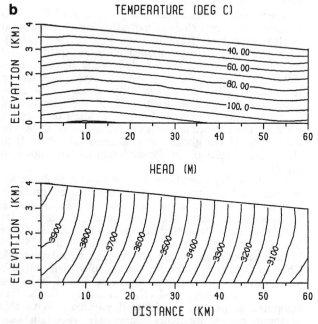

Fig.2b. The contour maps of T and h obtained in a forward solution of the simple geologic model illustrated in Fig.2a. The solution is used as the true model to be compared with models obtained by subsequent inversion.

TABLE 2. Material property values of the synthetic geologic model (refer to Fig.2 for the zone numbers).

κ zone number	κ value $(10^{-16}\ m^2)$	λ zone number	λ value $(W\ m^{-1}K^{-1})$
1	2.0	1	2.0
2	6.0		
3	3.0	2	2.5
4	0.6		

nodal points. Since the boundary flux at a boundary nodal point is grossly estimated from the vertical components of the flux values at the nearest numerical integration sample points and then assigned an appropriate standard deviation (STD), the components of the global nodal flux vector **q** related to the nodal points on this boundary always have uncertainties.

The results of inversion are presented in three ways:

1) Unnormalized PDF of λ and κ. This is used for the cases where the *a priori* information on the material properties is poor. The *a priori* and a posteriori PDF's and the true value of λ or κ are shown in the same plot. The improvement, if any, in our knowledge about the property is reflected by the changes in the expectation and the STD. The *a posteriori* PDF that has an expectation closer to the true value and an STD smaller than that of the *a priori* PDF, i.e., is narrower in shape, shows that our knowledge about the property has been improved.

2) Expectation contour map of T and h. This, and the next presentation method, is used for cases where the *a priori* information on the field variables is poor. Because there are too many temperature and head nodal values to be presented with PDF plots, we choose to contour the expectation values for each field variable.

3) STD contour map of T or h. Comparison of the *a posteriori* and the *a priori* STD contour maps shows the information gain at each point in terms of resolution. The a posteriori STD contour map itself shows in which parts of the cross-section a field variable is better resolved.

The STD and expectation contour maps reflect different aspects of the inversion results; a comparison of the *a priori* and the a posteriori contour maps gives an overall idea of how our knowledge of the uncertainty, as well as of the mean value, of a field variable is improved.

It should be noted that the STD contour map contours only the STD's of a field variable at nodal points, not the STD at all spatial points. The field variable value at a non-nodal point is interpolated using the basis functions and the nodal values of the element in which the point is located, as explained in the section on "Parameterization"; the variance of the value at this point should be obtained in a way consistent with the interpolation procedure, and will generally be different

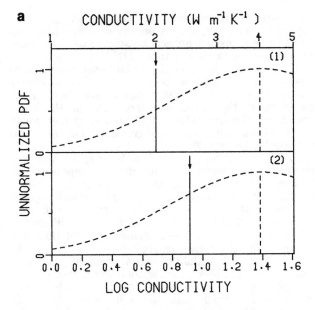

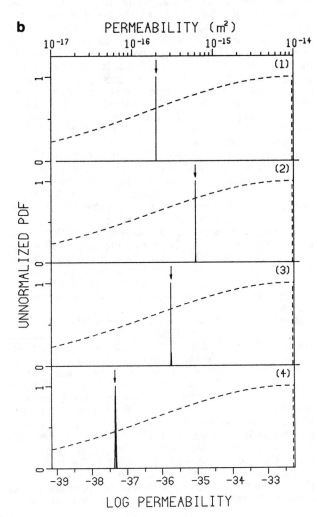

from that obtained by directly interpolating the nodal STD's as is done in most contouring computer programs. This, however, requires the values of the a posteriori covariances between nodal temperatures, which are not computed in this study for reasons mentioned earlier. However, even if the covariances were available, the computation of the STD at every point is unnecessary, because having obtained the true values at the nodal points (by forward calculation) a real test of the ability of the method to invert noisy data comes from a comparison of the *a priori* and the a posteriori values at these points; the nodal expectation and nodal STD contour maps fulfill this purpose.

Numerical Results

Case 1: Unknown material property values, almost noise free field variable data. Gaussian random noise N(0, 0.001 K) and N(0, 0.01 m) are added to the true values of nodal temperature and head. The STD's of T, 0.001 K, and of h, 0.01 m, are indeed very small, and hence tightly constrain the field variables. On the other hand, we assume that nothing is known about the material properties of the model. When we are totally ignorant of these parameters, the best we can do is to take a physically reasonable value as the *a priori* expectation for each material property zone, and assign to it a large STD, leaving the parameter effectively unconstrained. In this example, an erroneous uniform λ, and κ, is chosen as the *a priori* expectation for the whole medium with a large STD for its logarithmic transform (Fig.3). As can be seen in Fig.3b, the *a priori* expectations of permeabilities differ from the true values by orders of magnitude; however, because of the nearly "perfect" and well distributed field data, the geological model has been well resolved; that is, we have solved an ideal classical system identification problem.

Case 2: Very uncertain temperature values, almost noise free head and material property data. In this case, the nodal head values have Gaussian random noise of N(0, 0.0001 m), and a small STD of 0.001 is put on all the true logarithmic values of λ and κ. The true temperatures are perturbed with a Gaussian random noise N(0, 50 K), making the temperatures essentially unknown.

This example is, in a sense, an inverse solution to a slightly nonlinear forward problem, because with given material properties, known head values and specified boundary conditions, T can be readily obtained by a forward solution to equation (6), which is slightly nonlinear due to the temperature dependence of ρ and μ required to obtain **v**. This example shows how the

Fig.3. *A priori* (dashed lines) and *a posteriori* (solid lines) material property PDF's of case 1 with almost noise free temperature (STD = 0.001 K) and head (STD = 0.01 m) data. The numbers in brackets are the zone numbers. The true property value of each zone is indicated by an arrow.
(a) Thermal conductivity.
(b) Permeability.

knowledge of water movement serves to constrain the temperature field. It is well known that to obtain a unique forward solution to a potential field problem, the potential must be specified as a boundary condition at least at one point; if all the boundary conditions are given in term of specified fluxes (Neumann condition), there will be an infinite number of solutions for the potential field, differing from each other by a constant. If numerical methods are used, the resultant matrix to be inverted will be singular and no solution can be obtained. In this example, none of the nodal temperature values is well known *a priori*, but the accurate knowledge of the water flow pattern, which contains information on the temperature field through the temperature dependent parameters ρ and μ, strongly constrains the problem, and a unique solution is obtained as shown in Fig.4. The *a*

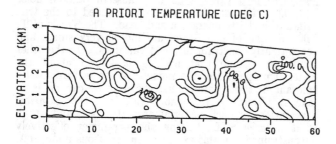

A PRIORI TEMPERATURE (DEG C)

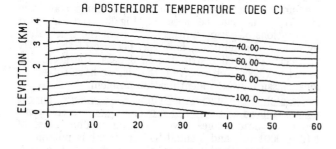

A POSTERIORI TEMPERATURE (DEG C)

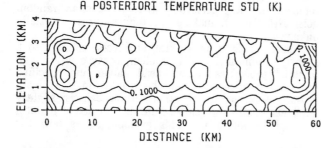

A POSTERIORI TEMPERATURE STD (K)

DISTANCE (KM)

Fig.4. The inverse solution of case 2 with almost noise free head and material property data but very uncertain temperature data (a weakly nonlinear problem in heat transfer). The *a priori* STD for nodal temperatures is a uniform 50 K. The contour interval for the *a priori* temperature is 50 °C. The contour maps show that the temperature field is resolved. Note, however, the concentric contours in the STD map, which is a result of nonlinearity.

posteriori STD of the nodal temperatures are, as can be expected, very small, but the pattern in the contour map needs to be explained further, because equal input noise has not resulted in equal output STD:

1) At higher elevation the nodal temperatures have larger STD. This is because the flow boundary conditions for the inverse problem are much more uncertain at the upper boundary than at the lower boundary. The uncertainties in the boundary flow inputs are translated into temperature gradients, and hence into the nodal temperatures.

2) There are some wave-like features in both the vertical and horizontal directions, appearing as "concentric" contours in the cross-section. These features show, in a way that is perhaps unique to inverse approaches, the response of the finite element numerical model to the nonlinear behavior of the partial differential equation (6). In a finite element formulation, the elemental conductivity matrix (14) depends on the coefficients of the partial differential equations, such as λ and $\mathbf{v}$ in (6), and hence, for the nonlinear problem in this example, depends on the temperature. Since the matrix is obtained through an integration over the whole element, using all the nodal temperature values of the element, the uncertainties in all the nodal temperatures are incorporated into the conductivity matrix. These uncertainties are in turn redistributed to individual nodal temperatures in the subsequent computations. Therefore, the temperature at a nodal point that is common to more elements will be influenced by the uncertainties in nodal temperatures at a larger number of nodes and, when the uncertainties in all the nodal temperatures are the same, have a larger STD. To illustrate this, we take as examples the nodal points at zero elevation, excluding the two end points. Referring to the finite element mesh in Fig.2a, we find that any node that is common to two elements has a larger STD thus forming a wave crest of the contour map on this line, while any node that belongs to only one element has a smaller STD thus forming a trough. Since the nonlinear term is proportional to the Darcian velocity, and the water flow is largely horizontal in the basin, the wave-like features of the STD map are more conspicuous in the horizontal direction than in the vertical direction. This phenomenon is characteristic of nonlinear problems; numerical results show that when a pure heat conduction problem, which is linear, is posed in exactly the same way as in this example, no wave-like features appear.

With noise STD = 50 K, a unique solution has been obtained. However, numerical experiments show that if the noise STD is increase to 100 K, iteration scheme (31) will not converge to the correct solution, due mainly to the restricted power of the scheme. As pointed out by Tarantola and Valette (1982), (31) is some kind of generalization of Newton's method for solving nonlinear systems. In fact, recalling that for a forward problem (24) the Jacobian matrix G is square and that $C_{ff} = 0$, one immediately finds that (31) reduces to Newton's iteration scheme. Newton's method generally requires the initial guess to be reasonably close to the true value. If this example is considered as a forward problem, then the *a priori* expectations of nodal temperature values serve as the initial guess, and should not be given too far

A PRIORI HEAD (M)

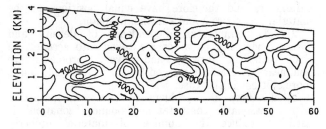

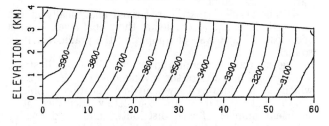

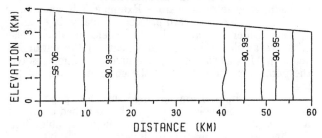

Fig.5. The inverse solution of case 3, in which the temperature and material property data are almost noise free (a linear problem in fluid flow). The a priori STD for nodal head values is a uniform 1000 m. The contour interval for the a priori head is 1000 m. The contour maps show that the head field is resolved.

from the true values, thus enforcing a limit on the noise level.

Case 3: Very uncertain head values, almost noise free temperature and material property data. This case is similar to the previous one except for the change of roles between temperature and head. Here the nodal temperature values are accurately known and well constrained (STD = 0.0001 K), while the nodal head values contain Gaussian random noise as large as N(0, 1000 m). The results are shown in Fig.5. In parallel to case 2, this example is a forward problem of equation (4) formulated in the inverse sense, with known temperatures as the constraints to the head values. Since no coefficients of equation (4) are head dependent, it becomes linear when the temperature field is given, and a solution can always be obtained in spite of the noise level in the a priori head data. Also, due to the linearity of the problem, the wave like features do not appear in the a posteriori STD contour map.

Case 4: Very uncertain material property values, noise in field variable data increasing with depth. The previous three examples have been highly idealized to illustrate some particular features of the inverse method. Here, we consider a more realistic case, in which the a priori uncertainties in both the material properties and the field variables are close to what may be expected in real data. The a priori information on λ and κ are shown by the PDF plots in Fig.6a and 6b; the a priori nodal values are obtained by perturbing the true nodal values of T and h with Gaussian random noise having STD's that increase with depth from a surface value 0.001 K to a bottom value 8.001 K for temperature, and 0.1 m to 40.1 m for head. The results of the inverse solution are presented in Fig.6.

Due to the noise in the temperature and head data, the material properties are not resolved so well as in case 1. The a posteriori STD's of λ and κ increase with depth because of the increasing noise in T and h with depth. In this example, the a posteriori STD patterns of the field variables are apparently determined by the a priori STD patterns, which can be clearly seen by a comparison of the a priori and the a posteriori STD contour maps of both nodal temperatures and heads (Fig.6c,d). Due to the nonlinearity of the problem, the wave like features can be observed in the a posteriori STD contour maps.

Summary and Discussion

A method to obtain a generalized inverse solution to the problem of 2–D coupled hydrological and thermal regimes has been presented. The method consists of two major steps: parameterization and parameter estimation. At the parameterization step, a 2–D finite element model is used, in which the field variables T and h are represented by their nodal values through the use of interpolation functions, and the material properties λ and κ are approximated by 2–D step functions, i.e., an average constant value over an element group is assigned to the group. The goodness of the zonation approach to the parameterization of material properties of complex subsurface structure depends inversely on the size of the zones. The partial differential equations relating T, h, λ and κ and their corresponding specified flow boundary conditions are then transformed into a nonlinear algebraic system relating the discretized values of these quantities and the boundary conditions. These discretized values form the parameter vector **p** which is to be estimated.

At the parameter estimation step, a Bayesian method is used, in which the a priori information on parameters is described by an a priori PDF of **p** and the above algebraic system is combined with the information on the flow boundary conditions to form a likelihood function, both of which are assumed to be jointly Gaussian. The most probable type Bayesian estimate is obtained by maximizing the a posteriori PDF of **p**, and a gradient method, namely the iteration scheme of Rodgers (1976), is employed for the maximization.

Our inverse method has the following major characteristics,

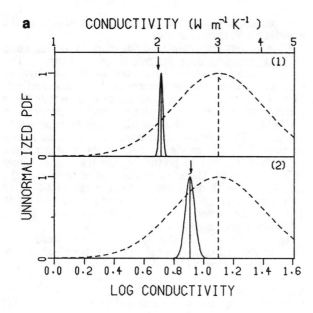

a

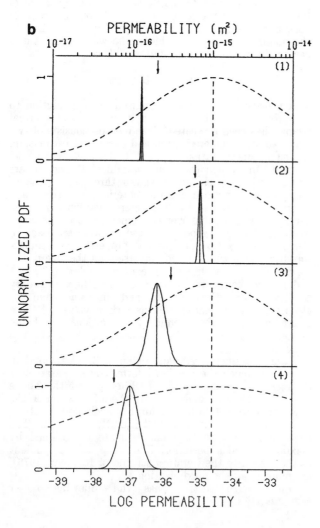

b

1) The field variables T and h are treated as parameters and therefore have equal positions to the material properties λ and κ in the inversion.

2) The temperature dependence of density and dynamic viscosity of water are taken into account, so that both the forward and inverse problems are nonlinear. A gradient method for the inverse solution is more efficient in this case.

3) To keep the problem of overparameterization within reasonable computational bound, "zonation" is used to reduce the number of material property parameters, and a quadrilateral isoparametric finite element model is used to reduce the number of field variable parameters.

4) The gradient matrix is derived analytically at the elemental level and assembled at the global level. This is the most efficient way of computing the gradient matrix so far.

The synthetic examples given in this work illustrate some general features of the inverse method. Example 1, which is an idealized system identification problem, shows that λ and κ are completely identified if a sufficient number of well distributed and good quality data are available on the field variables T and h. Example 2 shows how the uncertainties in the nodal values of the field variable interact in a nonlinear problem, inverse or forward. Example 3 is an inverse solution to a linear forward problem. Finally, in example 4, we consider a more realistic problem in which noisy data are available for all components of the parameter vector **p**. It was shown that the inverse method resolves the parameters provided that the noise level was not too high.

In the simple examples used in this paper, only Gaussian random noise, which can be described by known *a priori* PDF, is included in the field variable data. Further work is needed to investigate the effects of various other types of noise on the inverse solution, such as, the model error due to too few elements and nodal points and the error due to wrongly dividing material property zones. Reasonably well defined boundary conditions were applied to all the examples in this paper; cases involving very noisy, or uncertain boundary conditions are investigated in Wang and Beck (1989).

As Newton's iteration method is used in this work, only linear or mildly nonlinear problems can converge to a unique solution and retain quadratic convergency. This has been pointed out by Rodgers (1976), Tarantola and Valette (1982) and Wang and Beck (1987) and also observed in the numerical experiments of this work. To summarize the convergence behavior of our examples, we define convergence as when no component of the

Fig.6. The inverse solution of case 4, in which the material property values are very uncertain the STD's of noise in temperature and head data increase with depth.

(a) and (b) *A priori* (dashed lines) and *a posteriori* (solid lines) thermal conductivty and permeability PDF's showing that the properties are not as well resolved as in case 1 (Fig.3a) due to higher noise in tenperature and head data.

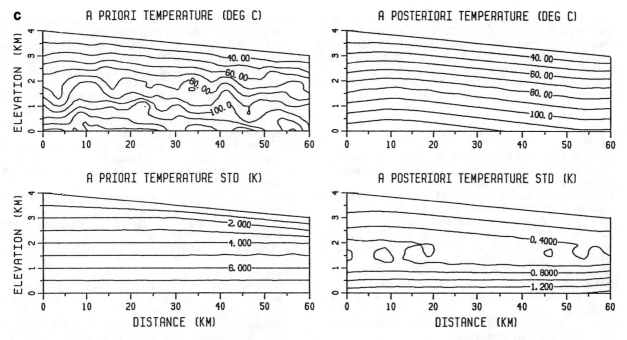

Fig.6c. The inverse solution of case 4: contour maps of nodal values and STD's for temperature.

parameter vector changes by more than 0.01 upon further integration. Example 3, which is an inverse solution to a linear problem, converges in two iterations (with a forward formulation, it should converge in one iteration); example 2, a slightly nonlinear problem which is ideal for Newton's method, converges in five iterations; example 1 is more nonlinear, and converges in eight iterations; finally, example 4, which is very nonlinear, converges in thirty–three iterations. Clearly, our numerical examples confirm the observation by Jackson and Matsu'ura (1985) that the nonlinearity of the inverse problem is effectively determined by the quality of the *a priori* information.

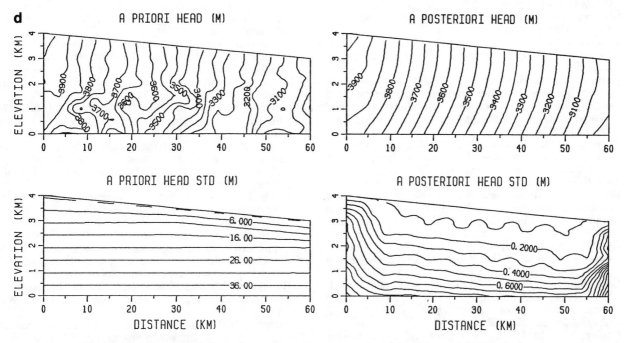

Fig.6d. The inverse solution of case 4: contour maps of nodal values and STD's for head.

Acknowledgements. The authors are in debt to A.D. Woodbury, S. Bachu, H. Pollack, S. Willett and S.B. Nielsen who read the manuscript and made many valuable suggestions. This research was carried out with the aid of a university research grant from Imperial Oil Limited and an operating grant from the National Sciences and Engineering Research Council of Canada.

References

Andrews, C.B, and Anderson, M.P., 1978. Impact of a power plant on the groundwater system of a wetland. *Ground Water*, 16(2), 105–111.

Andrews, C.B, and Anderson, M.P., 1979. Thermal alteration of groundwater by epage from a cooling lake. *Water Resour. Res.*, 15(3), 595–602.

Bachu, S., 1985. Influence of lithology and fluid flow on the temperature distribution in a sedimentary basin: a case study from the cold lake area, Alberta, Canada. *Tectonophysics*, 120, 257–284.

Bathe, K.J. and Wilson, E.L., 1976. *Numerical Methods in Finite Element Analysis*. Prentice–Hall, Englewood Cliffs.

Bear, J., 1972. *Dynamics of Fluids in Porous Media*. American Elsevier, New York.

Bejan, A., 1984. *Convection Heat Transfer*. Wiley, New York.

Beveridge, G.S.G. and Schechter, R.S., 1965. *Optimization: Theory and Practice*. McGraw–Hill, New York.

Box, G.E.P. and Tiao, G.C., 1973. *Bayesian Inference in Statistical Analysis*. Addison–Wesley, Reading, Massachusetts.

Burrus, J. and Bessis, F., 1986. Thermal modeling in the Provencal Basin (NW–Mediterranean), in J. Burrus (ed.): *Thermal Modeling in Sedimentary Basins*, Technip, Paris, pp. 393–416.

Burden, R.L., Faires, J.D. and Reynolds, A.C., 1981. *Numerical Analysis*. Prindle, Weber & Schmidt, Boston.

Carrera, J. and Neuman, S.P., 1986a. Estimation of aquifer parameters under transient and steady state conditions: 1. Maximum likelihood method incorporating priori information. *Water Resour. Res.*, 22(2), 199–210.

Carrera, J. and Neuman, S.P., 1986b. Estimation of aquifer parameters under transient and steady state conditions: 2. uniqueness, stability and solution algorithms. *Water Resour. Res.*, 22(2), 211–227.

Carrera, J. and Neuman, S.P., 1986c. Estimation of aquifer parameters under transient and steady state conditions: 3. application to synthetic and field data. *Water Resour. Res.*, 22(2), 228–242.

Cermak, V. and Jetel, J., 1985. Heat flow and ground water movement in the Bohemian Cretaceous Basin (Czechoslovakia). *J. Geodynamics*, 4, 285–303.

Chapman, D.S., Keho, T.H., Bauer, M.S. and Picard, M.D., 1984. Heat flow in the Uinta Basin determined from bottom hole temperature (BHT) data, *Geophysics*, 49: 453–466.

Cooley, R.L., 1977. A method of estimating parameters and assessing reliability for models of steady state groundwater flow: 1. theory and numerical properties. *Water Resour. Res.*, 13(2), 318–324.

Cooley, R.L., 1979. A method of estimating parameters and assessing reliability for models of steady state groundwater flow: 2. application of statistical analysis. *Water Resour. Res.*, 15(3), 603–617.

Dagan, G., 1986. Statistical theory of groundwater flow and transport: pore to labarotory, laboratory to formation and formation to regional scale. *Water Resour. Res.*, 22(9), 120S–134S.

Delfiner, P., 1976. Linear estimation of non stationary spatial phenomena, in *Advanced Geostatistics in the Mining Industry*, 49–68. D. Reidel, Dordrecht, Holland.

Delhomme, J.P., 1978. Kriging in hydrosciences. *Adv. in Water Resour.*, 1(5), 251–266.

Doligez, B., Bessis, F., Burrus, J., Ungerer, P. and Chénet, P.Y., 1986. in J. Burrus (ed.): *Thermal Modeling in Sedimentary Basins*, Technip, Paris, pp. 173–195.

Fradkin, L.J. and Dokter L.A., 1987. Statistical identification of hydrological distributed–parameter systems: theory and applications. *Water Resour. Res.*, 23(1), 15–31.

Freeze, R.A., 1975. A stochastic–conceptual analysis of one–dimensional groundwater flow in nonuniform homogeneous media. *Water Resour. Res.*, 11(5), 725–741.

Garven, G., 1985. The role of regional fluid flow in the genesis of the Pine Point deposit, Western Canada sedimentary basin, *Econ. Geol.*, 80: 307–324.

Garven, G., 1986. The role of regional fluid flow in the genesis of the Pine Point deposit, Western Canada sedimentary basin – A reply, *Econ. Geol.*, 81: 1015–1020.

Garven, G. and Freeze, R.A., 1984a. Theoretical analysis of the role of groundwater flow in the genesis of stratabound ore deposits: 1. Mathematical and numerical model. *Am. J. Sci.*, 284: 1085–1124.

Garven, G. and Freeze, R.A., 1984b. Theoretical analysis of the role of groundwater flow in the genesis of stratabound ore deposits: 2. Quantitative results. *Am. J. Sci.*, 284: 1125–1174.

Gavalas, G.R., Shah, P.C. and Seinfeld, J.H., 1976. Reservoir history matching by bayesian estimation. *Soc. Pet. Eng. J.*, 16, 337–350.

Gosnold, W.D. and Fischer, D.W., 1986. Heat flow studies in sedimentary basins, in J. Burrus (ed.): *Thermal Modeling in Sedimentary Basins*, Technip, Paris, pp. 199–217.

Hitchon, B., 1984. Geothermal gradients, hydrodynamics, and hydrocarbon occurrences, Alberta, Canada. *AAPG*, 66: 713–743.

Hoeksema, R.J. and Kitanidis, P.K., 1984. An application of the geostatistical approach to the inverse problem in two–dimensional groundwater modeling. *Water Resour. Res.*, 20(7), 1003–1020.

Huyakorn, P.S. and Pinder, G.F., 1983. *Computational Methods in Subsurface Flow*. Academic Press, New York.

Jackson, D.D. and Matsu'ura, M., 1985. A Bayesian approach to nonlinear inversion. *J. Geophys. Res.*, 90(B1), 581–591.

Jones, F.W., Majorowitz, J.A. and Lam, H.L., 1985. The variation of heat flow density with depth in the Prairies Basin of western Canada. *Tectonophysics*, 121, 35–44.

Kitanidis, P.K. and Vomvoris, E.G., 1983. A geostatistical approach to the inverse problem in groudwater modeling (steady state) and one–dimensional simulations. *Water Resour. Res.*, 19(3), 677–690.

Kasameyer, P., Younker, L. and Hanson, J., 1985. Inverse approach for thermal data from a convecting thermal system. *J. of Geodynamics*, 4, 165–181.

Loaiciga, H.A. and Marino, M.A., 1987. The inverse problem for confined Aquifer flow: identification and estimation with extensions. *Water Resour. Res.*, 23(1), 92–104.

Mercer, J.W. and Faust, C.R., 1980. Ground–water modeling: mathematical models. *Ground Water*, 18(3), 212–227.

Mercer, J.W., Pinder, G.F. and Donaldson, I.G., 1975. A Galerkin–finite element analysys of the hydrothermal system at Wairakei, New Zealand. *J. Geophys. Res.*, 80(17), 2608–2621.

Neuman, S.P. and Yakowitz, S., 1979. Astatistical approach to the inverse problem of aquifer hydrology: 1. theory. *Water Resour. Res.*, 15(4), 845–860.

Neuman, S.P. and fogg, G.E. and Jacobson, E.A., 1980. Astatistical approach to the inverse problem of aquifer hydrology: 2. Case study. *Water Resour. Res.*, 16(1), 33–58.

Rodgers, C.D., 1976. Retrieval of atmospheric temperature and composition from remote measurements of thermal radiation. *Rev. Geophys. Space Phys.*, 14, 609–624.

Roy, R.F, Beck, A.E. and Touloukian, Y.S., 1981. Thermophysical properties of rocks, in *Physical properties of rocks and minerals*, edited by Y.S. Touloukian, W.R. Judd and R.F. Roy. Mcgraw–Hill, New York.

Sauty, J.P., Gringarten, A.C., Menjoz, A. and Landel, P.A., 1982a. Sensible Energy Storage in Aquifers: 1. theoretical study. *Water Resour. Res.*, 18(2), 245–252.

Sauty, J.P., Gringarten, A.C., Menjoz, A. and Landel, P.A., 1982b. Sensible Energy Storage in Aquifers: 2. Field Experiments and comparisons with theoretical results. *Water Resour. Res.*, 18(2), 253–265.

Schweppe, F.C., 1973. *Uncertain Dynamic Systems*. Prentice–Hall, New Jersey.

Smith, L. and Chapman, D.S., 1983. On the thermal effects of groundwater flow: 1. regional scale systems. *J. Geophys. Res.*, 88(B1), 593–608.

Straus, J.M. and Schubert, G., 1977. Thermal convection of water in a porous medium: effects of temperature– and pressure–dependent thermodynamic and transport properties. *J. Geophys. Res.*, 82(2).

Sun, N.Z. and Yeh, W.W.G, 1985. Identification of parameter structure in groundwater inverse problem. *Water Resour. Res.*, 21(6), 869–883.

Tarantola, A., 1987. *Inverse Problem Theory*. Elsevier, New York.

Tarantola, A. and Valette, B., 1982. Generalized nonlinear inverse problems solved using the least squares criterion. *Rev. Geophys. Space Phys.*, 20(2), 219–232.

Vasseur, G., Lucazeau, F. and Bayer, R., 1986. The inverse problem of heat flow density determination from inaccurate data. *Tectonophysics*, 121, 25–34.

Wang, J.Y, Wang, J.A., Xiong, L.P. and Zhang, J.M., 1985. Analysis of factors affecting heat flow density determination in the Liaohe Basin, North China. *Tectonophysics*, 121, 63–78.

Wang, K. and Beck, A.E., 1987. Heat flow measurement in lacustrine or oceanic sediments without recording water bottom temperature variations. *J. Geophys. Res.*, 92(B12), 12837–12845.

Wang, K. and Beck, A.E., 1989. An inverse approach to heat flow study in hydrologically active areas. Geophys. *J. R. astr. Soc*, in press.

Willett, S.D. and Chapman, D.S., 1987. On the use of thermal data to resolve and delineate hydrologic flow system in sedimentary basins: an example from the Uinta Basin, Utah. In *Proceedings of the Third Canadian/American Conference on Hydrology of Sedimentary Basins: Application to Exploration and Exploitation*, edited by B. Hitchon, S. Bachu and C. Sauveplace, National Water Assn., Dublin, Ohio, 159–168.

Woodbury, A.D. and Smith, L., 1985. On the thermal effects of three–dimensional groundwater flow. *J. Geophys. Res.*, 90(B1), 759–767.

Woodbury, A.D., Smith, L., 1988. Simultaneous inversion of hydrogeologic and thermal data: 2. incorporation of thermal data. *Water Resour. Res.*, 24(3), 356–372.

Woodbury, A.D., Smith, L. and Dunbar, W.S., 1987. Simultaneous inversion of hydrogeologic and thermal data: 1. theory and application using hydraulic head data. *Water Resour. Res.*, 23(8), 1586–1606.

Yeh, W.W–G, 1986. Review of parameter identification procedures in groundwater hydrology: the inverse problem. *Water Resour. Res.*, 22(2), 95–108.

Yeh, W.W–G and Yoon, Y.S., 1981. Aquifer parameter identification with optimum dimension in parameterization. *Water Resour. Res.*, 17(3), 664–672.

Yeh, W.W–G and Yoon, Y.S. and Lee, K.S., 1983. Aquifer parameter identification with kriging and optimum parameterization. *Water Resour. Res.*, 19(1), 225–233.

Zienkiewicz, O.C., 1972. *The Finite Element Method in Engineering Science*. McGraw–Hill, New York.

USE OF DIMENSIONAL ANALYSIS IN THE STUDY OF THERMAL EFFECTS OF VARIOUS HYDROGEOLOGICAL REGIMES

Garth van der Kamp

Saskatchewan Research Council, Saskatoon, Saskatchewan S7N 2X8, Canada

Stefan Bachu

Alberta Research Council, Edmonton, Alberta T6H 5X2, Canada

Abstract. Actual cases of subsurface fluid-thermal interactions can very seldom be exactly quantified and analysed. The processes involved are complex and in most cases the available data for real systems are very limited. Dimensional analysis could be a useful tool for the description and analysis of fluid-thermal systems. This point has been recognized in the literature, but considerable confusion persists as to selection of the appropriate characteristic parameters.

Dimensional analysis of the basic energy transfer equation shows that, in general, two distinct Peclet numbers should be used to characterize horizontal and vertical heat transfer processes respectively. The relative magnitudes of horizontal convective heat transfer and vertical conductive heat transfer cannot be measured by a Peclet number alone, as is often done, but the relative magnitude of horizontal and vertical temperature differences must also be taken into account. The special case of a closed groundwater flow system, i.e. a system for which no fluid crosses the subsurface boundary, is of particular interest. For such a system, the nondimensional heat transfer equation takes a simpler form and a single Peclet number can be used as a measure of the effect of fluid motion on the subsurface thermal regime.

A number of case histories of fluid-thermal systems, taken from the literature, are analysed in terms of the appropriate Peclet number. In each case it is shown that the magnitude of this number characterizes the degree of thermal disturbance due to fluid movement.

The application of dimensional analysis to fluid-thermal systems can be useful at all stages of an investigation. In the exploration phase approximate calculations using Peclet numbers can serve to identify the dominant processes that are operating and to discern which thermal anomalies could be due to fluid flow. The insights gained in this manner can guide further data gathering. During numerical modelling the possible range of parameter values can be greatly decreased if appropriate values of Peclet numbers are taken into account.

Introduction

The geothermal regime in the Earth's crust is governed mainly by the transport to the surface of heat that flows outwards from the mantle and of heat that is generated internally by the decay of radioactive isotopes in the rocks. The main mechanisms of heat transfer are conduction and convection by moving fluids. In many cases the effect of convection is a priori assumed to be negligible, and the subsurface temperature distribution is calculated from the heat conduction equation. Of late, the effect of moving groundwater on subsurface heat transfer processes is recognized more and more to be an important element in earth sciences studies such as terrestrial heat flow, geothermal phenomena, hydrothermal circulation and deep waste disposal.

In geological environments characterized by the existence of permeable aquifers, fractured rocks, or faulted zones, there is movement of water driven by different processes and phenomena like gravity, compaction, buoyancy, capillarity and osmosis. Therefore, there is always a measure of convection in such an environment. The hydraulic and thermal properties of the rocks and the characteristics of the hydrogeological systems are such that, in some cases, conduction and convection of heat are of the same order of magnitude, while in other cases either one can be dominant. Therefore, it is important to understand both the hydrogeological and geothermal regimes in the subsurface environment.

Hydrogeological systems are usually very complex, and full analysis requires much data and generally sophisticated tools such as numerical models. In most cases, the necessary information and/or tools are not available, and the cost of acquiring them and processing the data is prohibitive in terms of time and/or money. In such cases, dimensional analysis can be used as a simple and effective tool in the assessment of the effects of groundwater flow on the subsurface thermal regime. Besides not requiring much information, dimensional analysis can actually help in devising data collection strategies by establishing which are the important processes in a system. While the usefulness of dimensional analysis has been recognized to a certain extent, some confusion persists with regard to the selection of the appropriate characteristic parameters. Use of inappropriate parameters may lead to mistaken conclusions, and it is therefore important that a solid theoretical basis is established for the parameters used in the analysis.

This paper presents the dimensional analysis theory as it applies to natural hydrogeological systems. The results are then applied to both theoretical and real cases described in the literature.

Analysis of Thermal Effects of Fluid Flow in Porous Media

Consider a representative element of a porous or fractured hydrogeological system (Fig. 1), for which it is possible to define equivalent porosity n and permeability k (Bear, 1972). The partial differential equation for heat transfer in a porous medium can be written as:

$$(\varrho c)_m \, \partial T/\partial t + n \varrho_f c_f \bar{v} \cdot \nabla T = \nabla \cdot \lambda_m \nabla T + S \qquad (1)$$

Copyright 1989 by
International Union of Geodesy and Geophysics
and American Geophysical Union.

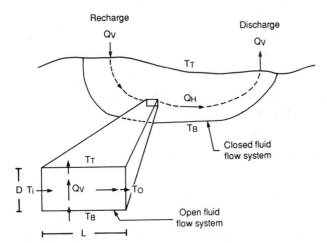

Fig. 1. Schematic representation of open and closed hydrogeological systems and of a representative element.

where T is temperature, $\bar{v}$ is the fluid velocity vector, S represents heat sources, ϱ is density, c is specific heat, λ is thermal conductivity and t is time. The subscripts f and s refer to the fluid and solid phases respectively, while the subscript m refers to the fluid-saturated porous medium. The heat capacity $(\varrho c)_m$ and the thermal conductivity λ_m of the porous medium are defined by:

$$(\varrho c)_m = n\varrho_f c_f + (1-n)\varrho_s c_s \tag{2}$$

$$\lambda_m = n\lambda_f + (1-n)\lambda_s \tag{3}$$

Expressions 2 and 3 are derived through analysis of heat transfer processes and averaging at the pore scale (Cheng, 1978; Bear and Bachmat, 1986).

In writing the energy equation 1, it is assumed that: the solid matrix is rigid (no fluidization of the porous bed), the viscous dissipation of energy is small and may be neglected, the adiabatic temperature changes are negligible, there is instantaneous thermal equilibrium between the solid and fluid phases at every point in the system, and that the thermal conductivity of the solid matrix is isotropic.

Because we are concerned here mainly with the overall thermal effects of various hydrogeological regimes, we can simplify equation 1 further by considering a system at steady state with no internal heat sources. The system is also taken to be homogeneous and isotropic with respect to thermal properties, but not necessarily with respect to the hydraulic ones. Moreover, for the sake of simplicity we consider a two-dimensional flow system with the x-axis oriented in the main horizontal flow direction. With these simplifying assumptions, equation 1 is reduced to:

$$n\varrho_f c_f \left(v_x \frac{\partial T}{\partial x} + v_z \frac{\partial T}{\partial z} \right) = \lambda_m \left(\frac{\partial^2 T}{\partial x^2} + \frac{\partial^2 T}{\partial z^2} \right) \tag{4}$$

where the subscripts x and z denote the respective components of the velocity vector. The more complex case of three-dimensional unsteady flow with heat sources can be treated in a similar form.

The representative element, taken as rectangular for easier mathematical treatment, is characterized by the following (Fig. 1): horizontal length L, height or thickness D, temperatures T_T at the top and T_B at the bottom, total horizontal fluid flow Q_H and total vertical fluid flow Q_V. The representative element can actually be the entire hydrogeological system, if the latter can be characterized by

equivalent properties in a basin as a whole. The components of the characteristic specific discharge vector are defined by:

$$q_H = Q_H/D \quad \text{and} \quad q_V = Q_V/L \tag{5}$$

In order to perform a dimensional analysis of the importance of convective versus conductive heat transfer in the representative element, the variables are made dimensionless as follows:

$$X = x/L; \quad Z = z/D; \quad \theta = (T - T_T)/(T_B - T_T); \tag{6}$$

$$q_X = nv_x/q_H; \quad q_Z = nv_z/q_V$$

By substituting equation 6 into equation 4, the latter becomes:

$$Pe^* \left[q_X \frac{\partial \theta}{\partial X} + \frac{Q_V}{Q_H} q_Z \frac{\partial \theta}{\partial Z} \right] = A^2 \frac{\partial^2 \theta}{\partial X^2} + \frac{\partial^2 \theta}{\partial Z^2} \tag{7}$$

where the dimensionless group:

$$Pe^* = \frac{\varrho_f c_f q_H D A}{\lambda_m} \equiv \frac{\varrho_f c_f Q_H A}{\lambda_m} \tag{8}$$

is defined as the "Geothermal" Peclet number, and A = D/L is the aspect ratio of the representative element.

The usual Peclet number for heat transfer, defined by:

$$Pe = v_0 l/\alpha \tag{9}$$

is the ratio of heat transferred convectively to heat transferred conductively in laminar flow far from any boundaries (Bennett and Myers, 1982). In the above expression, v_0 is a characteristic velocity, α is the thermal diffusivity of the fluid, and l is a characteristic length. The Peclet number for heat transfer in porous media (Bachu and Dagan, 1979) takes into account the Darcy flow through the pore space, and is defined as:

$$Pe' = n\beta \, v_0 l/\alpha_m \tag{10}$$

where $\alpha_m = \lambda_m/(\varrho c)_m$ is the thermal diffusivity of the fluid-rock system, and $\beta = \varrho_f c_f/(\varrho c)_m$ is the ratio of the heat capacity of the fluid to the heat capacity of the fluid saturated medium.

The Geothermal Peclet number Pe* takes into account also the geometric characteristics of a hydrogeological system defined by the aspect ratio A. For consistency and similarity with relations 9 and 10, the Geothermal Peclet number can be written also as:

$$Pe^* = \beta \frac{q_H D}{\alpha_m} A \equiv \beta \frac{Q_H A}{\alpha_m} \tag{11}$$

None of the Peclet numbers defined above is characterized by the existence of a critical value below which the system is conductive, as in the case of the critical Rayleigh number in natural convection (Neild, 1968; Rubin, 1975; Ribando and Torrance, 1976). However, for Peclet numbers much less than unity, the system is conduction dominated, while for values significantly greater than unity the system is convection dominated. An order of magnitude of one for a Peclet number indicates a conductive-convective system.

Application to Hydrogeological Systems

One of the most crucial aspects in dimensional analysis is the choice of the characteristic length l. The characteristic length depends, among other things, on the phenomena and the scale of the processes involved. For transport processes in porous media analysed at a microscopic scale, the characteristic length is the pore size or the grain size. However, here we deal with a much larger scale (mega to gigascopic), and the characteristic length has to be representative for the scale of the problem. The characteristic length is taken in the main direction of heat flux, which, in the case of terrestrial heat flow is vertical. Therefore, the characteristic length l becomes the thickness D of the hydrogeologic system (as shown by relations 8-11). The use of the horizontal length L of a hydrogeologic system as a characteristic length, instead of its thickness, brings only a gross overestimation of the importance of convective heat transfer, with a corresponding erroneous interpretation of the causes and effects in any geothermal pattern. A Peclet number based on the horizontal length L is in effect a measure of horizontal convective heat transfer relative to horizontal conductive heat transfer, and it does not provide a good measure of the thermal effects of fluid flow.

The value of the Geothermal Peclet number Pe^* can be underestimated in some cases by an improper choice of the characteristic horizontal length L. In a hydrogeological system, the value of L is given by the horizontal dimension of the flow path, and not by the geometric horizontal dimension of the system. For example, in the case of a sedimentary layer with a cellular flow pattern driven by buoyancy or by topography, the characteristic horizontal length L to be considered in Pe^* is the length of a flow cell, and not the length of the layer.

Another important aspect in the analysis of thermal effects of fluid flow in porous media is the proper choice of the characteristic velocity v_0 in expression 10. In the case of horizontal fluid flow only ($Q_V = 0$), the product nv_0 becomes the specific discharge q_H, such that:

$$Pe^* = Pe'A \qquad (12)$$

In the case of vertical fluid flow only ($Q_H = 0$), the product nv_0 in expression 10 becomes the specific discharge q_V. In this case, the mathematical formalism leads to:

$$Pe' = Pe^* \frac{Q_V}{Q_H} \qquad (13)$$

such that Pe' has to be used in equation 7.

Examination of equation 7 shows that all variables are of the order of unity, and the magnitude of the various terms is given by the magnitude of the respective dimensionless coefficients. In most cases of hydrogeological systems, except perhaps in mountain regions, the aspect ratio A is much less than unity, such that the horizontal conduction of heat can be neglected (i.e. first term on the right-hand side).

Hydrogeological systems can be classified generally as closed or open, depending on their boundaries and the type of inflow/outflow. Closed systems have no subsurface mass exchange with other systems (no-fluid flow boundaries), with the inflow/outflow being given by recharge/discharge (Fig. 1). An open system is characterized by subsurface inflow and outflow from or to other systems (Fig. 1). Any "cut" from a closed system is an open system.

In the case of a closed hydrogeological system, the total horizontal flow Q_H is equal to the total recharge or discharge Q_V. In this case, assuming that the aspect ratio A is much less than unity, equation 7 becomes:

$$Pe^* (q_x \frac{\partial \theta}{\partial X} + q_z \frac{\partial \theta}{\partial Z}) = \frac{\partial^2 \theta}{\partial Z^2} \qquad (14)$$

and the ratio of convective heat transfer to conductive heat transfer is given by the Geothermal Peclet number Pe^*.

In the case of an open hydrogeological system with only vertical fluid flow the characteristic velocity v_0 is the vertical velocity, there is no horizontal heat transfer, and equation 7 becomes:

$$Pe' \, q_z \frac{\partial \theta}{\partial Z} = \frac{\partial^2 \theta}{\partial Z^2} \qquad (15)$$

The ratio of convective versus conductive heat transfer in such a system is given by the Peclet number for heat transfer in porous media Pe'.

In the case of an open hydrogeological system with mainly horizontal fluid flow, the temperature T_i at the inflow is different from the temperature T_0 at the outflow, such that the term $\partial \theta / \partial x$ in equation 7 is no longer of the order of magnitude of unity as all the other terms. By neglecting horizontal conduction of heat and defining a horizontal dimensionless temperature $\theta' = (T - T_i)/T_0 - T_i)$, equation 7 is normalized to:

$$Pe^* \left[\frac{(T_0 - T_i)}{(T_B - T_T)} q_x \frac{\partial \theta'}{\partial X} + \frac{Q_V}{Q_H} q_z \frac{\partial \theta}{\partial Z} \right] = \frac{\partial^2 \theta}{\partial Z^2} \qquad (16)$$

In this case, the ratio of horizontal convective heat transfer to the vertical conductive heat transfer is given by:

$$Pe^* \left(\frac{T_0 - T_i}{T_B - T_T} \right) \qquad (17)$$

In the case of closed hydrogeological systems, it is more convenient to use the total flow Q instead of the specific discharge q in the computation of the Geothermal Peclet number, because it is more amenable to direct measurement (recharge rates, spring discharge), and because it is conceptually preferable to consider thermal effects as a function of flow, rather than of the interaction between permeability, fluid properties and hydraulic gradient which produces the flow. However, in the case of open systems, the Geothermal Peclet number usually has to be calculated using the specific discharge q determined according to Darcy's law. In such cases, the uncertainty in the characteristic values of certain hydraulic parameters, particularly permeability, can bring a large degree of uncertainty in the evaluation of thermal effects of various hydrogeological regimes.

Application of Dimensional Analysis

Analytical and Numerical Solutions

The application of dimensional analysis presented in the foregoing part of this paper is not a trivial exercise, as attested by the debate in the literature as to which forms of the Peclet number, if any, are appropriate for various hydrogeological systems. One of the most straightforward ways to verify that proper definitions of the Peclet number have been arrived at is to inspect available analytical solutions for some simple systems. These analytical solutions can generally be written in terms of dimensionless groupings of the system's physical parameters and one of these groupings should be the Peclet number for the system.

The simplest system is one with only vertical temperature gradients and fluid flow. The analytical solution for this case (Bredehoeft and Papadopulos, 1965) can be written in the form:

$$\theta(Z) = \frac{1 - e^{Pe'Z}}{1 - Pe'} \qquad (18)$$

where Pe' is defined by equation 10, and θ and Z are defined by relation 6. This form of the Peclet number is thus clearly the appropriate form for one-dimensional systems.

For a two-dimensional system the choice of the appropriate Peclet number is less obvious. The dimensional analysis which leads to equation 14 suggests that, for a simple closed system, analytical solutions can be written in terms of Pe*, dimensionless coordinates, and the aspect ratio A, if the latter is relatively large.

Domenico and Palciauskas (1973) have derived an analytical solution for forced convective heat transfer in regional groundwater flow. Their solution is only valid for the slightly perturbed case in which the temperature distribution differs only slightly from the purely conductive case. On the basis of their analysis, Domenico and Palciauskas propose a modified Peclet number which may be interpreted as "a relative measure of heat transfer by bulk motion of a fluid to conductive heat transfer in a regional porous body". A rewriting of their equations in the form of the present paper shows that their Peclet number Pe⁺ can be written as:

$$Pe^+ = \frac{L}{\pi D} Pe^* \qquad (19)$$

However, their analytical expression for the temperature distribution can be written in terms of Pe* as long as the aspect ratio, D/L, is small. For example, the temperature gradient at the ground surface (Z = 1) can be written as:

$$\left. \frac{\partial \theta}{\partial Z} \right|_{Z=1} = 1 + \frac{\pi}{2} Pe^* \cos(\pi X) \qquad (20)$$

where θ, X and Z are defined by relations 6. Similarly, the root-mean-square σ of the deviation of the heat flow at the ground surface from the basal heat flow HFD is given by:

$$\sigma = 1.1 \, Pe^* \, HFD \qquad (21)$$

These analytical results show clearly that for this two-dimensional case the disturbance of the temperature field by groundwater flow is characterized by Pe* as defined in this paper. The dimensionless group proposed by Domenico and Palciauskas corresponds to Pe⁺/π as may be seen by comparing the expression given in equation 19 with equation 12.

These analytical results indicate that Pe⁺, the Peclet number for flow in porous media as defined by relation 10, is appropriate for the case when fluid flow and conductive heat flow are parallel. However, for a two-dimensional hydrogeological system with mostly horizontal fluid flow and vertical conductive heat flow, Pe* rather than Pe⁺ gives a measure of the thermal disturbance due to convective heat transport. The analytical results thus confirm that the dimensional analysis presented in this paper provides a good characterization of subsurface fluid-thermal interactions.

The usefulness of dimensional analysis for cases of complex geometry can be evaluated by means of numerical modelling, such as the model studies described by Smith and Chapman (1983) and Woodbury and Smith (1985). These authors modelled closed groundwater flow systems and described the thermal effects of the groundwater flow. In particular, they characterized the perturbation of the near-surface thermal regime by means of the root mean square deviations of the near-surface heat flow from the basal heat flow.

Woodbury and Smith (1985) dealt with a three-dimensional system in which, however, a two-dimensional flow is dominant. They presented the results of their simulations in detailed descriptions, but Woodbury (1983) also summarized these same results in terms of recharge or total groundwater flow through the system and the deviation σ of the near-surface heat flow. These results, together with the

given geometry and thermal conductivity of the system, allow the Peclet number for each simulation to be calculated and compared with the corresponding heat-flow perturbation as measured by the deviation σ.

The Peclet number for each simulation was calculated from the data given by Woodbury (1983) by means of relation 8, with Q_H equal to the recharge per unit width of the system, L equal to 40 km, and D taken to be the mean depth to the bottom of the permeable part of the system (D takes values of 1.75 km to 4.75 km). The imposed uniform basal heat flux is 60 mW m⁻².

The results of this analysis are presented in Fig. 2 which shows on logarithmic scales the normalized deviation (σ/HFD) as a function of Pe*. Each point of the graph represents one of the model simulations described by Woodbury and Smith (1985). For details of each of these simulations the reader is referred to the appropriate section of their paper (the Woodbury and Smith figure number corresponding to each simulation is indicated on Fig. 2).

The deviation values used for Fig. 2 have been corrected for the deviation of the purely conductive case. This deviation is presumably related to topographic effects. It amounts to 0.8 mW m⁻², as shown by Woodbury's (1983) simulation with negligible permeability.

The major conclusion to be drawn from the results shown in Fig. 2 is that the geothermal Peclet number Pe* does indeed characterize the disturbance of the overall heat flow pattern as quantified by the deviation σ. Most of the plotted points lie on or near a straight line corresponding to:

$$\sigma = 1.3 \, Pe^* \, HFD \qquad (22)$$

where the imposed heat flux HFD at the base of the system is equal to 60 mW m⁻². This linear relationship is very similar to relation 20 for the two-dimensional analytical solution of Domenico and Palciauskas

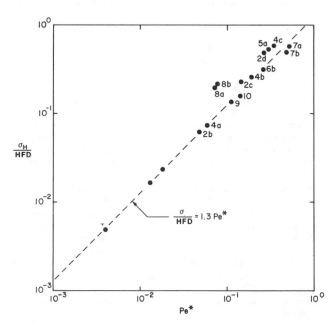

Fig. 2. Normalized deviation σ of surface heat flow (σ/HFD) as a function of Geothermal Peclet number Pe*. Points are identified by the corresponding figure number in Woodbury and Smith (1985). Unidentified points (from Woodbury, 1983) have the same system configuration as 2b and 4a, but with lower permeability.

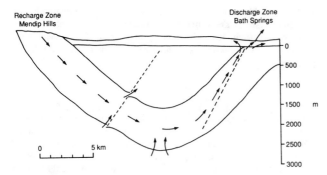

Fig. 3. Conceptual model of the flow path of thermal water in the Bath-Bristol Basin (after Andrews et al., 1982).

(1973). The slightly larger deviations σ for the numerical model are probably due to the somewhat three-dimensional nature of the flow system which imparts additional variability to the thermal regime as compared to the two-dimensional case.

Considering the range of system configurations represented in Fig. 2 the correlation between the deviation σ and Pe* is remarkably good. Clearly such a simple relationship between only two quantities can hardly be expected to describe the full range of heat flow variability for complex systems. Nor, as pointed out by Woodbury and Smith (1985), can such a simple relation describe the details of spatial variations of heat flow.

The linear relation described by equation 22 does allow further insight into the particular nature and cause of deviations from this overall pattern. In Fig. 2 the points labelled 8a and 8b for instance stand out as cases for which the deviation σ is twice as large as expected from the value of Pe*. The reason for this extra perturbation of the heat flow is easily identified. These two cases involve simulation of an aquifer situated at mid-depth in the system in which virtually all the horizontal fluid flow is concentrated. For these two cases, the natural choice for depth of the flow system is the bottom of the aquifer, because very little flow would penetrate deeper. The extra heat flow variability for 8a and 8b which shows up in Fig. 2 is therefore due to the fluid flow being concentrated at the bottom of the system. In contrast with the other cases, which are homogeneous, the average depth of fluid flow is only half the depth of the system, and the thermal perturbation due to the fluid flow is correspondingly less. The apparent anomaly of cases 8a and 8b would disappear if the depth D of the system, as used in the calculation of Pe*, were defined as the average depth of flow rather than as the full depth of the permeable part of the system. In practice, there will always be some question as to which value of depth to use, especially for complicated natural systems.

The foregoing case studies dealt with analytical and numerical models. The next step is to apply the results of dimensional analysis to cases of actual systems for which sufficient data are available to allow a verification and application of the dimensional analysis presented previously.

The Thermal Springs of Bath, England

Andrews et al. (1982) have given a detailed description of the thermal springs of Bath, England. A simplified sketch of the authors' conceptual model is given in Fig. 3. The reader is referred to the original paper for more detail. The thermal springs flow at a rate of 15 L s⁻¹ and a temperature of 46.5 °C. The authors hypothesize that recharge

probably takes place on the Mendip Hills, about 15 km away. The flow then descends to a depth of between 2700 and 4300 m, before relatively rapid ascent to the spring discharge points.

Heat flow in the basin is assumed to be average at 40 mW m⁻². The only geothermal gradient measured in the basin (11 mK m⁻¹) is known to have been affected by cold water inflows. The thermal yield of the springs above a non-thermal groundwater temperature of 10 °C is 2.2 MW.

A value for the geothermal Peclet number of the flow system can now be calculated. Andrews et al. (1982) suggest that the flow of the Bath springs derives from a 1 km wide segment of the aquifer zone, but this assumption may not be tenable. With a basal heat flow of 40 mW m⁻², the 2.2 MW thermal yield of the spring would take all the basal heat flow from 55 km² of the basin. With a length of flow path of 15 km, this energy balance consideration shows that the width of the flow path must be at least 3.7 km. Because there is certainly some conductive heat loss also, the width of the flow path can be estimated at 5 km. The flow per unit width then becomes 0.015/5000 = 3x10⁻⁶ m² s⁻¹. With a thermal conductivity of 2 W m⁻¹ K⁻¹, a length L of 15 km and a maximum depth for the flow system of 4.3 km, the resulting value for Pe* is 1.8. Considering the obviously three-dimensional nature of the flow, and the other assumptions that have entered into the calculation, the value of 1.8 must be considered a rough estimate only. Nevertheless, it provides quantitative insight into the hydrogeological and thermal processes which are operative.

First of all, a Geothermal Peclet number of about 1.8 implies strong disturbance of the subsurface temperature regime by groundwater flow. In the recharge end of the basin temperatures are likely to be well below normal, while in the discharge zone temperatures should be well above background. A Peclet number of the order of unity thus provides quantitative confirmation of the conceptual model for the origin of the thermal springs put forward by Andrews et al. (1982). The existence of thermal springs after all constitutes a direct indication of the elevated temperature in the discharge zone.

It should be emphasized that this is not a trivial result. If, for instance, the Peclet number had turned out to be much less than one, then subsurface temperatures would be near normal and thermal springs would be unlikely to occur. The conceptual model would then have to be revised. If, on the other hand, the Peclet number had turned out to be much greater than one, then the whole system would be strongly cooled by convection and the groundwater discharge would be relatively cool. (This is, in fact, the case for the 1 km wide flow path suggested by Andrews et al. which leads to Pe* = 9.0). In other words, thermal springs only occur for geothermal systems with a Peclet number of about unity, and any conceptual or quantitative model of such springs must be consistent with this criterion. The foregoing analysis of the Bath springs, using the Geothermal Peclet number combined with energy balance considerations, illustrates the value of such semi-quantitative analysis for evaluation of conceptual models.

Eastern Snake River Plain, Idaho, U.S.A.

Brott et al. (1981) in a study of heat flow in the Eastern Snake River Plain, Idaho, U.S.A., concluded that the average heat flow by fluid convection is about six times the average conductive heat flow. The data provided by the authors allow an estimate of Pe*. Taking D = 200 m, L = 400 km, Q_H = 4x10⁻³ m³ s⁻¹ and $λ_m$ = 2 W m⁻¹ K⁻¹, leads to Pe* = 4.2. This value of the Geothermal Peclet number for the system corresponds quite well to the measured ratio of convective to conductive heat flux.

It is of interest to note the different aspect ratios for the Bath-Bristol Basin and Eastern Snake River Plain. The values of A are

4.3/15 = 0.29 and 200/400,000 = 0.00005, respectively. A very "flat" system such as the Eastern Snake River Plain would normally show very little thermal disturbance, but in this case the flatness ($A = 5 \times 10^{-5}$) is compensated by the very high groundwater flow rates. The point that the Geothermal Peclet number Pe* provides a good measure of the thermal disturbance of these two very different natural systems can be considered as confirmation of the appropriateness and usefulness of this number for characterizing the thermal effects of various hydrogeological regimes.

Beaverhill Lake aquifer, Alberta, Canada

Bachu (1985) analysed the heat transfer processes in the Phanerozoic sequence of the sedimentary column in the Cold Lake area, Alberta. One of the aquifers, the Beaverhill Lake, is given here as an example of open systems. The aquifer is actually a carbonate platform which extends beyond the Cold Lake area, and is part of a wider regional flow system. In the area, the aquifer is underlain by a thick aquiclude (halite bed) and overlain mainly by a thick shaley aquitard. It dips to the southwest with a slope of about 4‰. The main hydrogeological and geothermal characteristics of the aquifer are: average thickness $D = 240$ m over a length $L = 150$ km, horizontal specific discharge $q_H = 3 \times 10^{-4}$ m/a, vertical specific discharge $q_V = 2 \times 10^{-7}$ m/a, and average geothermal gradient $\Gamma = 20$ mK m^{-1}. The fluid flow is updip to the northeast, with an average temperature at input of $T_i = 40\,°C$, and an average temperature at output of $T_O = 26\,°C$. In this case, the Geothermal Peclet number has a value of Pe* = 7×10^{-6}. According to relation 16, the ratio of horizontal heat transfer to the vertical conductive heat transfer is of the order of 2×10^{-5}, showing that the convection of heat by the formation waters is negligible in this case. The temperature difference of 14 °C between the input and output of the aquifer can be explained by purely conductive effects due to the difference in depth between the two, the input being deeper by about 600 m than the output.

Conclusions

Dimensional analysis is a simple and efficient tool which allows for the assessment of the relative importance of different factors and processes in a system without requiring its detailed knowledge. In the case of thermal effects of hydrogeological systems, the Peclet number provides a good measure of the importance of convective heat flow versus conductive heat flow and of the thermal disturbance produced by groundwater flow. Appropriate expressions of the Peclet number must be applied, depending on the type of hydrogeological system, and on the direction of the fluid flow with respect to the direction of the terrestrial heat flow. Improper use of various Peclet numbers or characteristic parameters of the hydrogeological system can lead to wrong conclusions and erroneous conceptual models of subsurface fluid flow and heat transfer processes. Conduction dominated hydrogeological systems are generally characterized by values of Peclet numbers less than 10^{-1}. Thermal springs occur for Peclet values around unity, while systems characterized by higher values are strongly cooled.

References

Andrews, J.N., Burgess, W.C., Edmunds, W.M., Kay, R.L.F. and Lee, D.J., The thermal springs of Bath. *Nature* 298:339-343. 1982.

Bachu, S., Influence of lithology and fluid flow on the temperature distribution in a sedimentary basin: a case study from the Cold Lake area, Alberta, Canada. *Tectonophysics,* 120:257-284. 1985.

Bachu, S. and Dagan, G., Stability of displacement of a cold fluid by a hot fluid in a porous medium. Phys. Fluids, 22(1):54-59. 1979.

Bear, J., *Dynamics of Fluids in Porous Media.* Elsevier, Amsterdam-New York, 764 pp. 1972.

Bear, J. and Bachmat, Y., Macroscopic modelling of transport phenomena in porous media. 2: Applications to mass, momentum and energy transport. *Transport in Porous Media,* 1(3):241-270. 1986.

Bennett, C.O. and Myers, J.E., *Momentum, Heat, and Mass Transfer.* McGraw Hill, 832 pp. 1982.

Bredehoeft, J.D. and Papadopulos, I.S., Rates of vertical groundwater movement estimated from the earth's thermal profile. *Water Resour. Res.,* 1:325-328. 1965.

Brott, C.A., Blackwell, D.D. and Ziagos, J.P., Thermal and tectonic implications of heat flow in the Eastern Snake River Plain, Idaho. *J. Geophys. Res.,* 86:11709-11734. 1981.

Cheng, P., Heat transfer in geothermal systems. *Advances in Heat Transfer,* 14:1-105. 1978.

Domenico, P.A. and Palciauskas, V.V. Theoretical analysis of forced convective heat transfer in regional ground-water flow. *Geol. Soc. Amer. Bull.,* 84:3803-3814. 1973.

Neild, D.A., Onset of thermohaline convection in a porous medium. *Water Resour. Res.* 4(3):553-560. 1968.

Ribando, R.J. and Torrance, K.E., Natural convection in a porous medium: effects of confinement, variable permeability and thermal boundary conditions. *J. Heat Transfer,* Trans. ASME, 98:42-48. 1976.

Rubin, H., On the analysis of cellular convection in porous media. *Int. J. Heat Mass Trans.,* 18:1483-1486. 1975.

Smith, L. and Chapman, D.S. On the thermal effects of groundwater flow, 1. Regional scale systems. *J. Geophys. Res.,* 88(B1):593-608. 1983.

Woodbury, A.D. *The thermal effects of three-dimensional groundwater flow.* M.Sc. thesis, University of British Columbia, Vancouver, B.C., 136 pp. 1983.

Woodbury, A.D. and Smith, L. On the thermal effects of three-dimensional groundwater flow. *J. Geophys. Res.,* 90:759-767. 1985.

TEMPERATURES, FLUID FLOW AND HEAT TRANSFER MECHANISMS IN THE UINTA BASIN

Sean D. Willet and David S. Chapman

Department of Geology and Geophysics, The University of Utah,
Salt Lake City, UT 84112, USA

Introduction

Both theoretical studies (Smith and Chapman, 1983; Garvin and Freeze, 1984a, 1984b; Luheshi and Jackson, 1986) and observational studies (Majorowicz and Jessop, 1981; Majorowicz et. al., 1985, 1986; Andrew-Speed et. al., 1984; Gosnold and Fischer, 1986; Willett and Chapman 1987a, 1987b) of sedimentary basin thermal regimes have shown that active groundwater systems can significantly perturb the temperature field that would exist if heat transfer within the basin were entirely conductive. The perturbation can affect hydrocarbon maturation in source rocks if the fluid flow regime and its accompanying lateral temperature variations exist for a sufficiently long time span. For these reasons it has become important to understand thermal processes in basins in greater detail.

In this extended abstract we summarize our thermal analysis of the Uinta Basin, western U.S.A. Details of our study are found in the following papers. A preliminary heat flow analysis of the Uinta Basin (Chapman et. al., 1984) had indicated significant lateral variations in the temperature and heat flow patterns. Subsequent and more detailed analyses (Willett and Chapman 1987a, 1987b) have confirmed the laterally varying thermal field, and have shown that the temperature perturbations are probably caused by a regional, topographically driven groundwater flow system. Most recently the analysis was extended to include implications of the thermal analysis for the thermal history of the basin and hydrocarbon maturation (Willett and Chapman, 1987c).

The Uinta Basin and its Thermal Regime

The Uinta Basin is an intermontane sedimentary basin within the northern Colorado Plateau of the U.S.A. The basin is roughly elliptical, measuring 210 kilometers along the major east-west axis and 160 kilometers in a north-south direction. Rapid subsidence began in early Tertiary; up to 6000 meters of sediment were deposited between 60 and 30 Ma. Oil is produced from the early Tertiary Wasatch and Green River formations. These are overlain by the late Eocene Uinta Formation, an interfingering unit of fluvial and lacustrine sediments, and at the surface by the Duchesne River

Formation. The entire Uinta Basin has experienced regional uplift and erosion in the last 10 My. The modern drainage system has provided about 1000 m. of elevation difference between the center of the basin and the north and south flanks.

The temperature field in the Uinta Basin was determined by analyzing 320 bottom hole temperatures (BHTs) in 270 wells whose positions are shown in Fig. 1. All BHTs used were corrected for the thermal disturbance of drilling and mud circulation by the Horner Plot method or a statistical correction based on the depth and shut-in time of the well (Willett and Chapman, 1987b). To obtain the best estimate of the temperature field throughout the basin, a method of stochastic inversion was used (Willett and Chapman, 1987a, 1987b). The thermal field obtained from the inversion is parameterized by smoothly varying temperature gradients within each formation. An example of this solution is given in Fig. 1, which shows temperature gradients in the Uinta Formation, one of the four Tertiary formations present in the Uinta Basin. Temperature gradients in the formation vary from 15 mK m^{-1} to over 35 mK m^{-1}. The temperature at any point can be calculated from the thickness and gradient maps of each formation in the basin.

The most notable thermal feature in these gradient maps is the low gradient region at the northwest extent of data coverage and the increase in gradients towards the geographic center of the basin along the Duchesne and Green Rivers. To illustrate better this lateral heterogeneity of the temperature field we calculated the temperature along a profile trending approximately perpendicular to the temperature gradient contours and the basin structure (Fig. 1). The temperature along this section is shown in Fig. 2a superimposed on the Tertiary stratigraphy. Temperatures at the base of the Tertiary section vary from 80 to 120 °C. This figure also shows a significant pull-down in the isotherms at the north end of the cross-section, reflecting the decrease in temperature gradients.

This lateral variation of temperature can be seen more clearly represented as a temperature anomaly. For example, in Fig. 2b a temperature anomaly has been calculated by subtracting the temperature field corresponding to the average temperature gradient (25 mK m^{-1}) from the observed field of Fig. 2a. The low gradients in the north side of the basin result in a negative temperature anomaly with a maximum of -25 °C at the base of the Green River Formation. There is also a broader positive anomaly in the center of the basin, or south end of the cross-section, that is barely discernible as a gradual up-warping of the isotherms in Fig. 2a.

Copyright 1989 by
International Union of Geodesy and Geophysics
and American Geophysical Union.

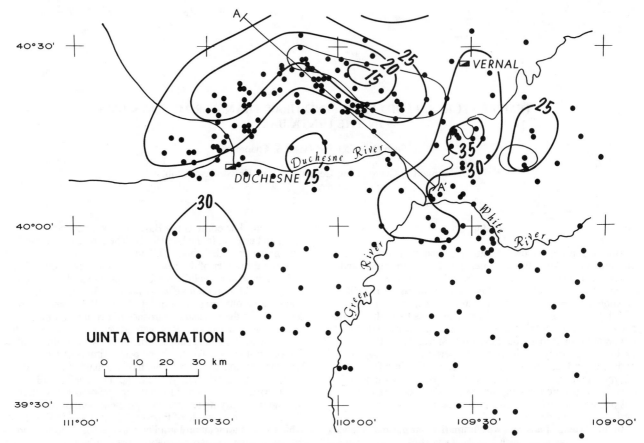

Fig. 1. Location map for thermal analysis of the Uinta Basin, northern Colorado Plateau, Utah, U.S.A. Drillholes with bottom hole temperatures (BHTs) shown by solid dots. Contours indicate map of thermal gradients within the Uinta Formation deduced from inversion of BHT data; contours in mK m^{-1}. Profile A-A' shows location of model section.

Interpretation of the Temperature Field

Temperature anomalies of this magnitude are most likely the result of either variations in the thermal conductivity of the sediment fill and surrounding rocks, an active groundwater flow system or a combination of both. Another possible mechanism is a variation in the basal heat flow into the basin, but it seems unlikely that there would be such a short wavelength variation in basal heat flow given the thickness of sediment cover and the consistency of heat flow measurements to the north (Bauer and Chapman, 1986; Deming and Chapman, 1988) and to the south (Bodell and Chapman, 1982). To determine which processes may be affecting the thermal state of the Uinta Basin we calculated a series of forward models with estimated thermal and hydrologic parameters (Willett and Chapman, 1987a, 1987b). A finite element code was used to calculate the coupled heat transfer and fluid flow for a specified thermal conductivity and permeability structure (Smith and Chapman, 1983); the corresponding temperature anomaly was then compared to the observed anomaly for model calibration. Because the profile used for the models was chosen to be parallel to the topographic (and presumably hydraulic) gradient as well as

perpendicular to the basin structure, a 2-D assumption for fluid and heat transfer is reasonable.

Thermal conductivity variations within the Uinta Basin and surrounding rocks produce a temperature anomaly with some similarities to the observed anomaly (Fig. 2b) but fail to produce the magnitude of the anomaly (Willett and Chapman, 1987b). Models with the thermal conductivities for the sediment fill given by Chapman et al.(1984) and Willett and Chapman (1987b) and the conductivity of the surrounding sediments given by Bauer and Chapman (1986), and assuming no advective heat transfer, resulted in a maximum negative anomaly of -12 °C at the base of the Green River Formation (see Fig. 15 of Willett and Chapman, 1987b). To account for the additional negative temperature anomaly we consider the hypothesis of thermal perturbation by groundwater flow.Willett and Chapman (1987a) discussed the possible groundwater systems in a semi-quantitative fashion, and suggested that the most probable flow system in the basin is in the relatively shallow, unconfined aquifer consisting of the Duchesne River Formation and the upper Uinta Formation. If this aquifer has a permeability of 5×10^{-15} m^2, the combined effects of fluid flow and heterogeneous thermal conductivity produce a temperature field with an

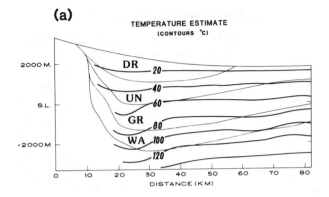

(a)

TEMPERATURE ESTIMATE
(CONTOURS °C)

(b) TEMPERATURE ANOMALY
(CONTOURS °C)

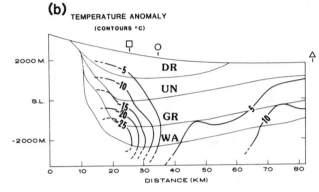

(c) TEMPERATURE ANOMALY
(CONTOURS °C)

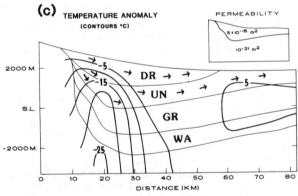

anomaly shown in Fig. 2c. This anomalous field is quite close to the observed anomaly of Fig. 2b. The maximum amplitude of the negative anomaly in the Tertiary section is almost 25 °C compared to the observed anomaly of just over 25 °C, a difference that is within the noise level. Additional simulations indicate that model permeabilities cannot be changed by more than a factor of ten from this value without seriously misfitting the temperature field.

Direct constraints on permeabilities are more difficult to obtain. Permeability measurements of the aquifer rocks are limited in number and spatial distribution. Furthermore, available data are difficult to interpret because of the scale-dependent nature of permeability (Brace, 1984). The Duchesne River and Uinta formations, which comprise the aquifer, consist of a series of interbedded lacustrine shales and sandstones. The interconnectivity of these beds is not well known, nor are the facies changes that affect lateral permeability changes. The lateral permeability of the aquifer is determined by the permeability of the sands and their continuity and connectedness. Vertical permeability is determined more by the permeability of the flow-inhibiting shales, their lack of continuity and the degree of fracturing. These factors are all difficult to account for in estimating permeability. Laboratory measurements at the core scale or even pump tests at a larger scale do not sample spatial heterogeneities, and therefore laboratory and pump tests must be used with caution when inferring permeabilities appropriate for a basin scale model.

Measured permeabilities, however, do provide some constraint on forward modeling. In this case they probably represent a minimum value of the regional average horizontal permeability. Figure 3 shows a compilation of permeability

Fig. 2. (a) Temperature field for section A-A' in Fig. 1 constructed from local formation gradients. Isotherms contoured at 20 °C interval. Geometry of the Tertiary units of the Uinta Basin is shown by thin lines; symbols refer to formations: Duchesne River (DR), Uinta (UN), Green River (GR), Wasatch (WA). (b) Uniform gradient temperature anomaly for section A-A' in Fig. 1. This anomaly is the difference between the temperature field of Fig. 2a and the temperature field that would result if the thermal gradient were everywhere 25 mK m⁻¹. Anomaly contours in °C. Symbols above section represent location of wells where vitrinite reflectance results are available. (c) Uniform gradient temperature anomaly along A-A' for a model simulating both conductive and advective heat transfer. Thermal conductivity varies within and between formations and was constrained by measurements (see text and references for details). Permeability is shown in inset; flow in this simulation is restricted to the Duchesne River and Uinta formations.

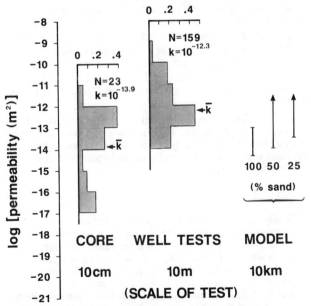

Fig. 3. A comparison of Duchesne River Formation permeabilities: left - laboratory measurements on core samples; center - well tests; right - model permeabilities required to produce the observed temperature anomaly. Model value varies depending on the effective thickness of the permeable beds within the formation. As thickness, (i.e. % sand) is decreased the minimum permeability required to produce an equivalent thermal effect increases.

measurements on 23 core samples and 159 single well pump tests in the Duchesne River Formation. These data were compiled by the United States Geological Survey and the Utah Department of Natural Resources (Hood, 1976). Not included in this figure are two more accurate multi-well tests whose mean lies very close to the mean of the single well tests. The mean permeability of the core tests is $10^{-13.9}$ m². The mean of the pump tests is $10^{-12.3}$ m², more than an order of magnitude higher, indicating the significance of spatial heterogeneity. The minimum permeability from the model lies between $10^{-14.3}$ m² and $10^{-13.5}$ m². We believe that the model permeability is lower than measured permeabilities because of a sampling bias in the measurements. Primarily sandstone aquifers are selected for testing, and the measurements reflect the local horizontal permeability of these rocks. In contrast, the model permeability is a regional average which samples the integrated effects of sandstones interbedded with lower permeability shales. This effect is shown by the two open interval model permeabilities labeled 50% and 25% sand in Figure (3). These represent the results for models with anisotropic permeability. In this case the temperature effects are limited by the vertical permeability (shales) and the horizontal permeability (sands), is not constrained by the temperature field.

An additional constraint on the flow system and so indirectly on the aquifer permeability is provided by the total water budget. Only a limited portion of the total precipitation is available as recharge to any groundwater system. An estimate of the water available for recharge in the northern Uinta Basin is provided by Hood and Fields (1978). They assume 10% of all precipitation is recharged into the groundwater system. Of this, approximately 40% is discharged by local scale flow systems. The total precipitation in the northern drainage basin is 6.0×10^9 m³/yr giving a total of 3.6×10^8 m³/yr available for regional groundwater recharge. The average vertical velocity in the modeled profile is 1.5×10^{-10} m s⁻¹. Assuming discharge is taking place at elevations under 2000 m. gives a total discharge area of 2300 km². Integrating over this area gives a volume flux of 1.1×10^7 m³/yr. Thus the model requires much less water than is actually available. The discrepancy is easily accounted for by either more discharge at a local scale or additional regional scale flow at greater depths as hypothesised by Willett and Chapman (1987a).

Thermal History

Finally we have investigated the likely duration of the asymmetric thermal field in the Uinta Basin by examining thermal history indicators. As data constraints for our analysis we use vitrinite reflectance determinations provided by Chevron U.S.A., Northern Region, from three wells (Fig. 4) on or near the cross section discussed previously: Ute Tribal 1-21-72, Chenney A2 1-33, and Red Channel Federal 22-1. These three wells, projected along strike of the thermal field to positions shown in Fig. 2b, constitute a fortuitously good test of our deduced temperature field, as they occur (a) in the center of the negative temperature anomaly, (b) on the edge of the pronounced negative temperature anomaly, and (c) in the positive temperature anomaly respectively. Figure 4 shows vitrinite reflectance values plotted with depth for these wells. Reflectance values at a given depth are generally lowest

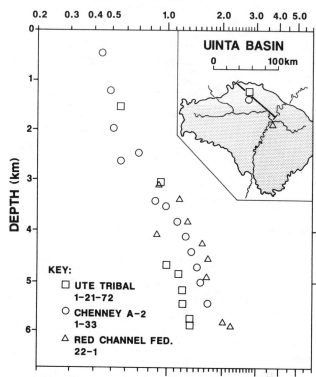

Fig. 4. Vitrinite reflectance versus depth for 3 wells along profile A-A' in Fig. 1. Locations of wells are shown here in inset and on Fig. 2b in relation to present day temperature anomaly.

for Ute Tribal 1-21-72, the northernmost well, intermediate for Chenney A2 1-33, and highest for Red Channel Federal 22-1, the southernmost of the three available wells. This pattern is qualitatively consistent with the present thermal state of the basin along the cross section (Fig. 2a,b), provided the three sites have experienced a similar burial history. To investigate quantitatively the detailed thermal history we have modeled time-temperature histories for the three wells under two contrasting assumptions: (1) that geothermal gradients have been a uniform 25 mK m⁻¹ in all formations throughout the basin's evolution, and (2) that the present laterally variable gradients in each formation have existed for the history of the basin. This test is not intended to predict the absolute thermal history of the basin which would require data or assumptions regarding the heat flow history and compaction of the sediments, but rather is intended to show that the relative difference in thermal state across the basin has existed for much of the history of the basin. Results indicate that the hypothesis of a constant gradient through time fails; the assumption of the present formation gradients with their lateral variation having existed throughout the Uinta Basin history provides a better explanation of the vitrinite reflectance data (Willett and Chapman, 1987c).

Our model of a laterally varying temperature field extending back through time to explain the vitrinite reflectance data contrasts with a maturation model for the same basin described recently by Sweeney et. al. (1986, 1987) and

Burnham and Braun (1985). In applying their detailed chemical model they assumed a constant thermal gradient of 25 mK m^{-1} for the entire basin. Although their model correctly predicted the maturation for the Altamont- Bluebell field on the boundary of our depressed gradient region, they underestimate the maturity levels in the Redwash area, a region of enhanced present day thermal gradients. They attribute the discrepancy to oil migration from a more mature source area. Our results suggest that the difference may be due to hydrologically induced lateral heat flow variations.

Conclusions

We demonstrate that careful statistical treatment of bottom hole temperatures gives an estimate of the thermal state of a basin with sufficient accuracy to allow detailed forward modeling of the thermal processes controlling that field. In the Uinta Basin we find a large lateral variation in temperature gradients that produces lateral temperature differences of up to 35 °C at the base of the Tertiary section. This temperature variation is consistent with the conductivity structure of the basin with a thermal overprint resulting from advective heat transport by an active groundwater flow system. Vitrinite reflectance data support the hypothesis that these temperature variations have existed through the history of the basin.

Acknowledgments. Acknowledgment is made to the donors of the Petroleum Research Fund, administered by the American Chemical Society, for partial support of this research. Support was also received from Chevron Oil Field Research Company, LaHabra, and ARCO Oil and Gas Company. Vitrinite reflectance data were provided by Chevron U.S.A., Northern Region, Denver. Much of the research reported in this paper was completed while one of us (DSC) was at the Pacific Geoscience Centre, Sidney, Canada; this paper is contribution No. 19688 of the Geological Survey of Canada.

References

Andrews-Speed, C. P., E. R. Oxburgh, , and B. A. Cooper, Temperature and depth-dependent heat flow in western North Sea, Am. Assoc. Petrol. Geol. Bull., 68, 1764-1781, 1984.

Bauer, M.S., and D.S. Chapman, Thermal regime at the Upper Stillwater Dam site, Uinta Mountains, Utah: Implications for terrain, microclimate and structural corrections in heat flow studies, Tectonophysics, 128, 1-20, 1986.

Bodell, J. M., and D. S. Chapman, Heat flow in the north-central Colorado Plateau, J. Geophys.Res., 87, 2869-2884, 1982.

Brace, W., Permeability of crystalline rocks: new in situ measurements, J. Geophys. Res., 70, 4327-4330, 1984.

Burnham, A. K., and R. L. Braun, General kinetic model of oil shale pyrolysis, In Situ, 9, 1- 23, 1985.

Chapman, D. S., T. H. Keho, M. S. Bauer, and M. D. Picard, Heat flow in the Uinta Basin determined from bottom hole temperature (BHT) data, Geophys., 49, 453-466, 1984.

Deming, D. and D. S. Chapman, Inversion of bottom-hole temperature data: the Pineview field, Utah-Wyoming thrust belt, Geophys., 53, 707-720, 1988.

Garvin, G., and R. A. Freeze, Theoretical analysis of the role of groundwater flow in the genesis of stratabound ore deposits. 1. Mathematical and numerical model, Am. J. Sci., 284, 1085-1124, 1984a.

Garvin, G., and R. A. Freeze, Theoretical analysis of the role of groundwater flow in the genesis of stratabound ore deposits. 2. Quantitative results, Am. J. Sci., 284, 1125-1174, 1984b.

Gosnold, W. D., and D. W. Fischer, Heat flow studies in sedimentary basins, in Thermal Modeling in Sedimentary Basins, edited by J. Burrus, 199-217, Editions Technip, Paris, 1986.

Hood, J. W., Characteristics of aquifers in the northern Uinta Basin Area, Utah and Colorado,Tech. Publ. No. 53, State of Utah Department of Natural Resources, 1976.

Hood, J. W., and F. K. Fields, Water resources of the northern Uinta Basin Area, Utah and Colorado, with special emphasis on ground-water supply, Tech. Publ. No. 62, State of Utah Department of Natural Resources, 1978.

Luheshi, M. N., and D. Jackson, Conductive and convective heat transfer in sedimentary basins, in Thermal Modeling in Sedimentary Basins, edited by J. Burrus, 219-234, Editions Technip, Paris, 1986.

Majorowicz, J. A., and A. M. Jessop, Regional heat flow patterns in the western Canadian sedimentary basin, Tectonophysics, 74, 209-238, 1981

Majorowicz, J. A., F. W. Jones, H. L. Lam, and A. M. Jessop, Regional variations of heat flow differences with depth in Alberta, Canada, Geophys. J. R. Astron. Soc., 81, 479-487, 1985.

Majorowicz, J. A., F. W. Jones, and A. M. Jessop, Geothermics of the Williston Basin in Canada in relation to hydrodynamics and hydrocarbon occurrences, Geophys., 51, 767-779, 1986.

Smith, L., and D. S. Chapman, On the thermal effects of groundwater flow. 1. Regional scale systems, J. Geophys. Res., 88, 593-608, 1983.

Sweeney, J. J., A. K. Burnham, and R. L. Braun, A model of hydrocarbon maturation in the Uinta Basin, Utah, U.S.A., in Thermal Modeling in Sedimentary Basins, edited by J. Burrus, 547-561, Editions Technip, Paris, 1986.

Sweeney, J. J., A. K. Burnham, and R. L. Braun, A model of hydrocarbon generation from type Ikerogen: application to Uinta Basin, Am. Assoc. Petrol. Geol. Bull., 71, 1967-985, 1987.

Willett, S. D., and D. S. Chapman, On the use of thermal data to resolve and delineate hydrologic flow systems in sedimentary basins: an example from the Uinta Basin, Utah, in Proceedings of the Third annual Canadian/American Conference on Hydrogeology. - Hydrology of Sedimentasry Basins: Application to Exploration and Exploitation, edited by B. Hitchon, S. Bachu, and C. Sauveplane, 159-168, Dublin, Ohio, 1987a.

Willett, S. D., and D. S. Chapman, Analysis of temperatures and thermal processes in the Uinta Basin, in Sedimentary Basins and Basin-Forming Mechanisms edited by C. Beaumont and A.J. Tankard, Can. Soc. Petrol. Geol. Memoir 12, 447-461, 1987b.

Willett, S. D., and D. S. Chapman, Temperatures, fluid flow and the thermal history of the Uinta Basin, in Migration of Hydrocarbons in Sedimentary Basins, edited by B. Doligez, 533-552, Editions Technip, Paris, 1987c.

HYDROLOGIC CONSTRAINTS ON THE THERMAL EVOLUTION OF THE RHINE GRABEN

Mark Person and Grant Garven

Department of Earth and Planetary Sciences,
The Johns Hopkins University, Baltimore, MD 21218

Abstract. The role of subsurface fluid flow in the thermal evolution of the Rhine Graben is investigated using transient finite element modeling. The analysis helps elucidate the importance of convective-heat transfer on the thermal maturation of organic matter in rift sediments and permits comparison of the different driving forces on fluids during basin development. Modeling results indicate that in Oligocene time, during a period of marine incursion, groundwater flow was driven by sediment compaction. Heat transfer during this time period was dominated by conduction, as groundwater velocities were too small to disturb the temperature field. Uplift of the Graben shoulders in conjunction with marine regression during Miocene to Recent times probably initiated a regional gravity-driven groundwater flow system. Convective-heat transfer during this period resulted in high heat flow (greater than $120 \ mW \, m^{-2}$) within groundwater discharge areas near the center of the rift and low heat flow (less than $20 \ mW \, m^{-2}$) within recharge areas near fault escarpments. Computed petroleum maturation indices (TTI) indicate that oil generation occurred much deeper in regions of groundwater recharge than in discharge areas. The model results are in good agreement with observed geothermic data and occurrences of petroleum within the rift.

Introduction

In continental rifts, circulating fluids can have an important effect on heat transport and thermal maturation of organic material in sediments during basin development. Field investigations [Stallman, 1963; Lewis and Beck; 1977; Majorowicz and Jessop, 1981; and Chapman et al., 1981] and theoretical studies [Bredehoeft and Papadopulos, 1965; Garven and Freeze, 1982, 1984a,b; and Smith and Chapman, 1983] have shown that advective heat transport can produce significant deviations from a purely conductive thermal regime under favorable hydrologic conditions within sedimentary basins. Unfortunately, the general pattern of deep groundwater flow within continental rift basins is known in only a few cases, and only qualitatively, as most previous studies were based solely on the interpretation of geochemical and geothermal data [Craig, 1966, 1969; Bentor, 1969; Manheim, 1974; Issar, 1979; Morgan, et al., 1981]. The detailed hydrodynamics of flow in rifts is virtually unknown, and yet there is clearly a need to understand the regional fluid flow if the thermal history of these basins is to be fully understood.

This paper seeks to investigate the role of paleogroundwater flow on the thermal history of one well-known continental rift: the Rhine Graben. A two-dimensional finite element model is presented that depicts the hydrothermal evolution of this rift system over the last 40 Myr. Petroleum maturation indices are calculated during basin development to illustrate the impact of convective-heat transfer on the thermal maturation of organic material in sediments. The two-dimensional simulations permit evaluation of the relative importance of the different driving forces on fluids during basin formation. The model results are compared with observed temperature patterns and paleo-heatflow data collected within the rift.

Geologic Setting

The Rhine Graben is one of the most studied continental rift systems in the world. The Graben, located in northwest France and southeast Germany, is about 300 km long and 40 km wide with syn-rift sediment accumulations as great as 3 km in some places (Fig. 1). Geophysical studies indicate that the lithosphere underlying

Copyright 1989 by
International Union of Geodesy and Geophysics
and American Geophysical Union.

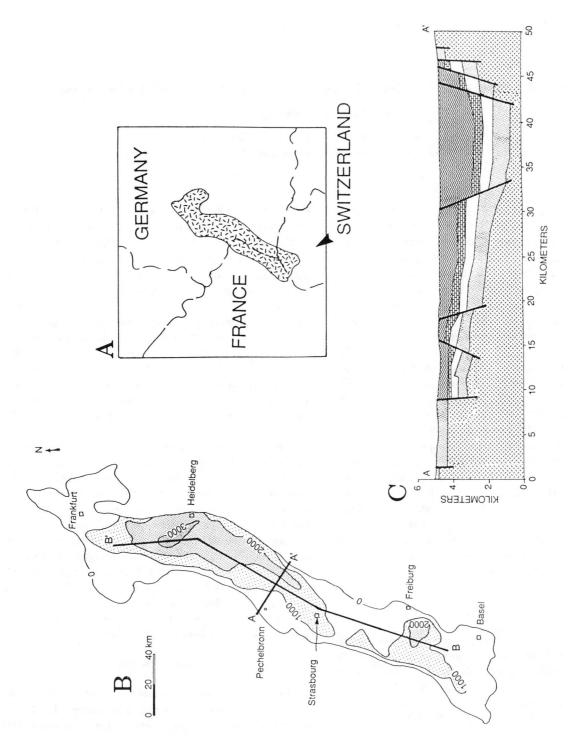

Fig. 1. (A) Location of the Rhine Graben within northwest France and southeast Germany. (B) Thickness of Cenozoic sediments within Rhine Graben (contour interval 1000 m) and location of the cross sections discussed in the paper [after Doebl et. al, 1974]. (C) Hydrogeologic cross section through the Rhine Graben, Pechelbronn Subbasin, along A − A'. Descriptions of the hydrostratigraphic units depicted in Figure 1c are presented in Table 1 and Figure 7.

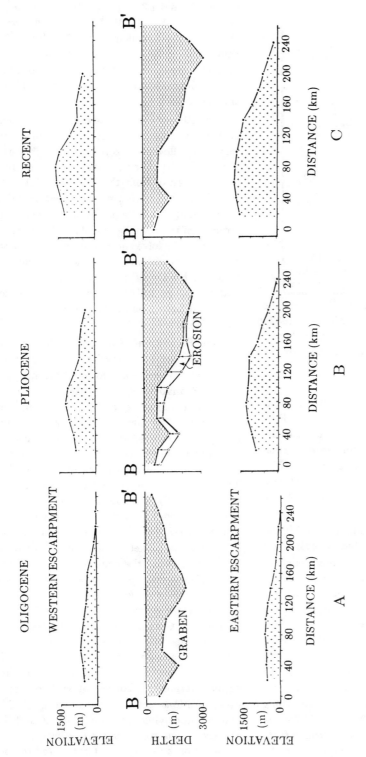

Fig. 2. Schematic diagram illustrating the tectonic evolution of the Rhine Graben and fault escarpments along the section $B - B'$ (see Figure 1b for the location of section $B - B'$). Sediment thicknesses and elevation of fault escarpments represent averaged values. Subsidence rate along the rift was greatest in the southern depocenter during Oligocene time (Fig 2a). However, the basin underwent uplift and erosion of the southern depocenter in Pliocene time (Fig. 2b). Subsidence in the northern depocenter has dominated during the Pliocene to Recent with rift-sediment thicknesses exceeding 3000 m (Fig. 2c). Uplift of the escarpment has continued gradually over the history of the rift [after Villemin et al., 1986].

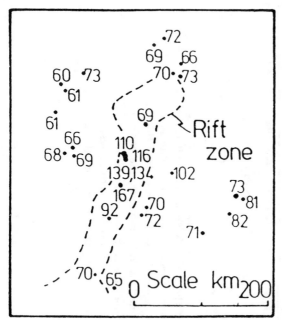

Fig. 3. Observed heat flow (in mWm^{-2}) within the Rhine Graben and adjacent areas [from Morgan, 1982]. Heat flow is highest in the middle segment of the rift and decreases to background levels in areas adjacent to the Graben.

the Rhine Graben has been thinned by stretching, necking, and associated upwelling of the asthenosphere [Edel et al., 1975; Zucca, 1984]. The basin is characterized by a long sediment-filled depression cut by numerous normal faults which trend parallel to the axis of extension (Fig. 1c). The Rhine Graben is thought to have formed as a result of tensional and sinistral forces associated with the Alpine Orogeny [Illies; 1970, 1978; Celal Sengor et al., 1978; and Villemin et al., 1986]. Beginning in Middle Eocene time, extensional stresses resulted in fault-block subsidence and lacustrine sedimentation along depocenters in the north and south [Illies, 1978; Illies and Greiner 1978] (see Fig. 2a). During the Oligocene, some segments of the rift underwent rapid subsidence as the Graben was flooded by a shallow intercontinental sea. Basin tectonism began to change in Middle Miocene to Early Pliocene with marine regression and the cessation of subsidence in the southern portion of the rift [Illies, 1978]. During the Middle to Late Pliocene time, sinistral forces acting along the axis of the Graben resulted in uplift and erosion along the southern portion of the rift (Fig. 2b). Roll [1979] estimates that about 500 m of sediments were eroded from the southern depocenter. Subsidence has continued in the northern depocenter into Recent time with sediments being derived from the southern portion of the rift (Fig. 2c).

Mean subsidence rates in the northern Rhine Graben today are between 3 to 7 $\times 10^{-5} myr^{-1}$ [Groten et al., 1979] and between 1 to 4 $\times 10^{-5} myr^{-1}$ in the southern segment [Malzerand and Schlemmer, 1975]. Roll [1979] has established that uplift of the rift escarpment over the last 40 million years has been gradual and continuous with maximum topographic relief between the fault escarpment and the valley floor occurring today.

The Rhine Graben, like many rift basins, is characterized by high geothermal gradients [Morgan, 1982]. Observed heat flow reaches 167 $mW m^{-2}$ in the center portion of the rift (Fig. 3) and decreases to background levels (60 $mW m^{-2}$) away from the Graben. Subsurface temperature data indicate that positive geothermal anomalies are found mostly in the center of the Graben along the middle segment of the rift (Fig. 4). Areas near the Graben shoulders generally have cooler temperatures than areas in the center of the basin. Analysis of temperature data collected within the rift indicates that the numerous geothermal anomalies are partly the result of forced convection caused by deep groundwater flow [Clauser; 1987 and Otto and Tóth; 1988].

The Pechelbronn subbasin within the Rhine Graben was chosen for our numerical study because of hydrogeologic and thermal data available from earlier investigations. Schnabele et al. [1948] has described in detail the petroleum geology of the Pechelbronn subbasin. Otto and Tóth [1988] have collected and analyzed hydraulic head, temperature, and geochemical data near Pechelbronn to study the association between groundwater flow patterns and the occurrence of hydrocarbon reservoirs. Their study has shown that the Pechelbronn oil fields occur in areas of groundwater discharge, which supports the hydraulic theory of petroleum entrapment of Tóth [1980]. Clauser [1987] developed and applied a three-dimensional hydrothermal model of the Rhine Graben near Manheim in order to study the conductive and convective components of heat transfer. Clauser [1987] calculated thermal Peclet numbers ranging between 0.08 and 1.30, suggesting important convective heatflow effects in some areas. Person and Garven [1987] used a steady-state finite element analysis to characterize the present-day hydrothermal system along the cross-section $A - A'$ (Fig. 1b). The goal of the steady-state modeling was to select a group of aquifer parameters consistent with observed hydrothermal conditions and to determine the relative importance of different driving-forces on fluids. Results indicated that gravity-driven groundwater flow was the dominant driving force on fluids and that computed temperatures were more sensitive to variations in hydraulic parameters (over two standard deviations) rather than thermal variables.

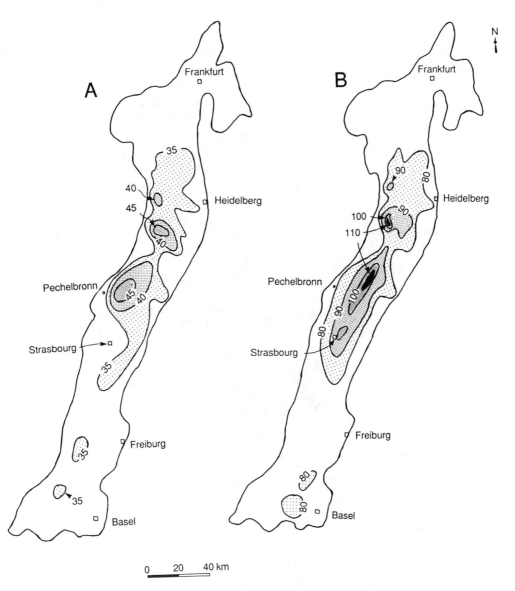

Fig. 4. Observed subsurface temperatures (°C) at 500 m (A) and 1500 m (B) below sea-level [after Haenel, 1979]. Highest temperatures occur near center of the rift valley along the middle segment of the Graben.

The model cross-section A − A' is situated between two sediment depositional centers intersecting the Pechelbronn oil fields near the eastern margin of the rift (Fig. 5). The occurrence of petroleum within the Rhine Graben also appears to be associated with areas of high heat flow. Most oil fields are located within the middle segment of the rift and many of them occur along faults that are discharging geothermal waters [Otto and Tóth, 1988].

We believe that present-day fluid migration within the Rhine Graben has components of flow parallel and per-

pendicular to the axis of the rift. The water-table configuration along the Rhine Graben in the central portion of the rift (Fig. 6) indicates that along-axis groundwater flow is an important component of flow in the south. However, flow perpendicular to the axis appears to dominate in the center and northern segments of the rift. Across A − A' the water table elevations vary 300 m from the center of the rift to the top of the fault escarpments. One goal of this paper is to present a methodology for quantifying the transient hydrology of the rift, so as to understand

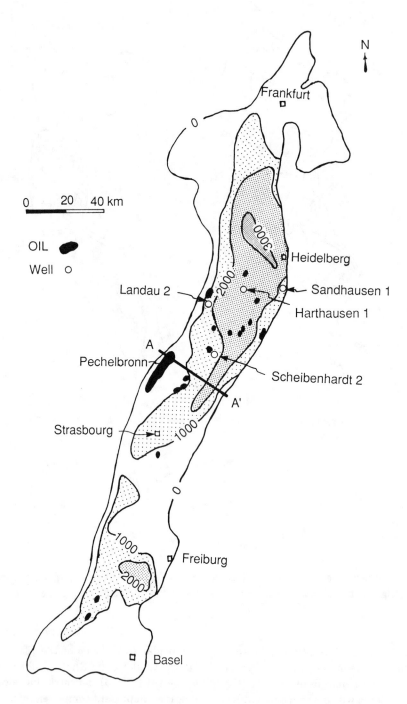

Fig. 5. Thickness of Cenozoic sediments within Rhine Graben (contour interval 1000 m), location of wells used by Espiltalie [1984] to estimate paleogeothermal gradients, and occurrence of petroleum within the rift [after Doebl et. al, 1974; Espitalie, 1984].

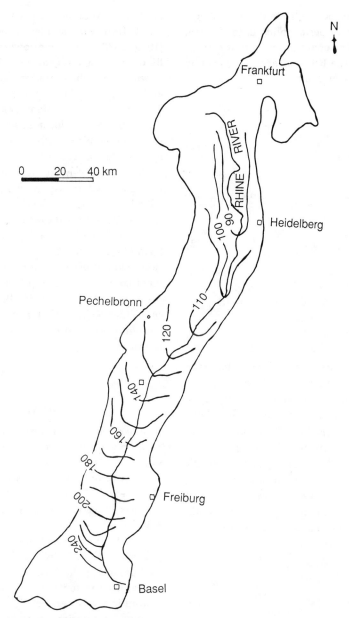

Fig. 6. The watertable elevation (contour interval 10 m) within the Rhine Graben. Watertable configuration indicates components of flow parallel and perpendicular to the long-axis of the rift, although mostly transverse flow occurs in the northern section. [after Karrenberg and Struckmeier, 1970].

better the present-day flow patterns and their influence on heat transport.

Mathematical Model of Transport Processes in Rifts

In continental rift basins, subsurface fluids can be driven by spatial variations in fluid density, by sediment compaction, and by gravitational forces. During basin development, temporal variations of hydrologic and thermal boundary conditions could change the relative importance of these different driving mechanisms. Fluid flow and heat transfer within a sedimentary basin can be mathematically modeled following a continuum approach, although limitations always exist with regard to representing fractured rocks, scale effects of local heterogeneity, and the availability of permeability data [Garven and Freeze, 1982, 1984a]. The equations used to represent transport processes within the Rhine Graben are given by the following conservation equations:

Solid Mass:

$$\frac{\partial V_z}{\partial z} = \left(\frac{1}{1-\phi}\right)\frac{\partial \phi}{\partial t} \tag{1}$$

Fluid Mass:

$$\nabla\left[\rho_f \mu_r \frac{\rho_o g k}{\mu_o}(\nabla h + \rho_r \nabla Z)\right] = \rho_f \phi \rho_o g \beta\left(\frac{\partial h}{\partial t} + V_z\right)$$

$$+ \left(\frac{\rho_f}{1-\phi}\right)\frac{\partial \phi}{\partial t} - \rho_f \phi \alpha_t \frac{\partial T}{\partial t} \tag{2}$$

Thermal Energy:

$$\nabla \cdot [\lambda \nabla T] - \rho_f C_f \bar{q} \cdot \nabla T = [\rho_f C_f \phi + (1-\phi)\rho_s C_s]\frac{\partial T}{\partial t}$$

$$+ \left(\frac{\rho_f h_f}{1-\phi}\right)\frac{\partial \phi}{\partial t} \tag{3}$$

The symbols are defined as follows: velocity of the porous media V_z, elevation above datum Z, porosity ϕ, Darcy velocity q, intrinsic permeability k, relative viscosity $\mu_r = \frac{\mu_o}{\mu}$, reference fluid viscosity μ_o, fluid viscosity μ, hydraulic head h, relative fluid density $\rho_r = \frac{\rho - \rho_o}{\rho_o}$, reference fluid density ρ_o, fluid density ρ, gravitational constant g, fluid compressibility coefficient β, thermal expansivity coefficient of the fluid α_t, bulk thermal conduction-dispersion tensor λ, fluid temperature T, specific heat capacity of fluid C_s and solid C_f, and fluid enthalpy h_f. The bulk thermal conduction-dispersion is a tensor given by:

$$\lambda_{xx} = \lambda_f^\phi \lambda_s^{1-\phi} + \rho_f C_f \phi \frac{\epsilon_l q_x^2 + \epsilon_t q_z^2}{|q|} \tag{4}$$

$$\lambda_{zz} = \lambda_f^\phi \lambda_s^{1-\phi} + \rho_f C_f \phi \frac{\epsilon_l q_x^2 + \epsilon_l q_z^2}{|q|} \tag{5}$$

$$\lambda_{xz} = \rho_f C_f \phi \frac{(\epsilon_l - \epsilon_t)q_x q_z}{|q|} \tag{6}$$

with bulk thermal conductivity in the x direction λ_{xx}, bulk thermal conductivity in the z direction λ_{zz}, off diagonal component of the bulk thermal conductivity λ_{xz}, thermal conductivity of the rock λ_s, thermal conductivity of the fluid λ_f, the Darcy velocity of the fluid in the x and z direction q_x and q_z respectively, magnitude of Darcy velocity $|q|$, longitudinal dispersivity ϵ_l, and transverse dispersivity ϵ_t. Since the thermal conductivity of the solids is several times greater than the fluid, the bulk-thermal conductivity will increase as porosity decreases. Equations

(1) to (3) were derived assuming a Lagrangian coordinate framework similar to the approach taken by Bethke [1985, 1986]. The equations are valid for inhomogeneous fluid flow in heterogeneous and anisotropic porous media in which the solid grains are assumed to be incompressible and the fluid and solid phases are in thermal equilibrium. The system of equations can account for the effects of sediment compaction and gravity-driven flow. Equation (3) accounts for the effects of heat transfer by conduction and forced convection.

The finite element method was used to solve the fluid flow and heat transport equations. Triangular elements with linear interpolation functions were used to approximate unknown hydraulic heads, fluid velocities, and temperatures [Huyakorn and Pinder, 1983]. A finite difference form of equation (1) was used to solve for the consolidation velocity of the porous medium in response to sediment compaction. The porosity versus depth relations as presented by Bethke [1985, 1986] were used to predict the consolidation rate that appears in equations (1) to (3). His empirical relations for porosity and permeability take the form of:

$$\phi = \phi_o exp(-Bz) \tag{7}$$

$$log(k) = -C + D\phi \tag{8}$$

where ϕ_o, B, C, and D are functions of the porous media. Equations (1) to (3) were solved in a sequential manner as the simulation marched forward in time. Because porosity is taken to be a function of depth and hence the velocity of the solids V_z, equation (1) is nonlinear and three to four iterations were required to arrive at stable values for ϕ and V_z.

The boundary conditions used for sediment compaction (Equation 1) included specified velocities of the porous medium along the bottom row of nodes and fixed elevation along the top row of nodes. For groundwater flow, hydraulic heads were specified along the land-surface and no flux boundaries were the assigned to the sides and base of the finite element mesh. Between 40 and 15 Myr ago, hydraulic heads along the land-surface were held constant at $4670\ m$ to represent marine incursion. During the last 15 Myr, hydraulic heads near the edges of the basin were allowed to increase in an attempt to represent marine regression and uplift of the Graben framing faults. Because the position of the top row of nodes was not varied during basin formation, hydraulic heads greater than land-surface elevation had to be specified for theses nodes during this period of subareal exposure. This assumption introduces little error since the variation in water table elevation is small compared to the total thickness of the flow system [Tóth, 1962]. The boundary conditions for the thermal submodel include specified temperature along the top row of nodes, specified heat flux along the bottom row of nodes, and insulated sidewalls. Basal heat flux was varied through time in order to approximate the lithosphere thinning and associated increase in geothermal gradient. The heat flux was allowed to increase linearly from 60 mWm^{-2} to present day values estimated by Clauser [pers. comm.] to be around 80 mWm^{-2}.

The thermal maturity of organic matter is a useful indicator of heat flow conditions during sedimentary basin evolution [Lopatin, 1971] and maturation is frequently used as an indicator for the onset of oil generation. The thermal maturation of organic material within sediments of the Rhine Graben was calculated in our model using the time-temperature kinetic equation presented by Waples [1980]:

$$TTI = \int_0^t a_1^{(a_2T-a_3)} dt \qquad (9)$$

where TTI is the petroleum maturation index, and a_1, a_2, and a_3 are empirical temperature-dependent coefficients. Inspection of equation (4) indicates that petroleum maturation increases exponentially with temperature, but is linear in time. Waples [1980] indicates that petroleum generation occurs within organic rich beds over a range of TTI values between 15 and 160.

Hydrothermal Model Results

A transient model of the Rhine Graben was constructed along the section $A - A'$ in order to study the hydrother-

mal evolution of this continental rift basin. The values for the coefficients ϕ_o and B were estimated based on values given in Bethke [1985, 1986] and from porosity-depth data for the Graben presented in Roll [1979]. Permeability-porosity coefficients C and D were estimated from parameters used by Person and Garven [1987]. Description of rock types and formation ages for the hydrostratigraphic units are listed in Figure 7. Rock properties used in the model for the different hydrostratigraphic units are listed in Table 1. Hydrostratigraphic unit 1 is composed of low permeability ($3 \times 10^{-14} cm^2$) granites, migmatites, and gneisses of Paleozoic age. The crystalline rocks are overlain by the principal aquifer, unit 2, which is comprised of Permian sandstone and carbonate rocks. The permeability assigned to unit 2 ranged from 3 to $6 \times 10^{-10} cm^2$. Hydrostratigraphic unit 3 is made up of low permeability (8×10^{-12} to $8 \times 10^{-11} cm^2$) clays and marls of Triassic age. The aquitard is overlain by a thin dolomitic aquifer of Eocene age (unit 4). The dolomite permeability varied between $5 \times 10^{-11} cm^2$ to $3 \times 10^{-10} cm^2$. The entire section is capped by a thick sequence of mostly low permeability clays and marls with some interbedded sand, and gravel deposits of Oligocene to Quaternary age (unit 5). Permeabilities assigned to Unit 5 ranged from 8×10^{-12} to $8 \times 10^{-11} cm^2$.

The hydrostratigraphic development of the rift, as calculated by the model, is shown at 10 million year intervals in Figure 8. The average rate of basin subsidence was set at $3 \times 10^{-5} myr^{-1}$. The kinematics of lystric normal fault movements could not be represented by the model since equation (1) permits subsidence only in the z direction along nodal columns. However, representation of

TABLE 1. Rock Properties

Hydrostratigraphic Unit	Porosity-Depth Coefficients		Permeability-Porosity Coefficients		Thermal Conductivity
	ϕ_o	B	C	D	λ_s $(Wm^{-1}K^{-1})$
5	0.60	6.0×10^{-04}	17.0	5.0	2.5
4	0.55	5.5×10^{-04}	15.0	4.0	2.5
3	0.60	6.0×10^{-04}	17.0	5.0	2.5
2	0.50	5.0×10^{-04}	13.0	2.0	2.5
1	0.05	0.0×10^{-04}	22.0	4.0	2.5

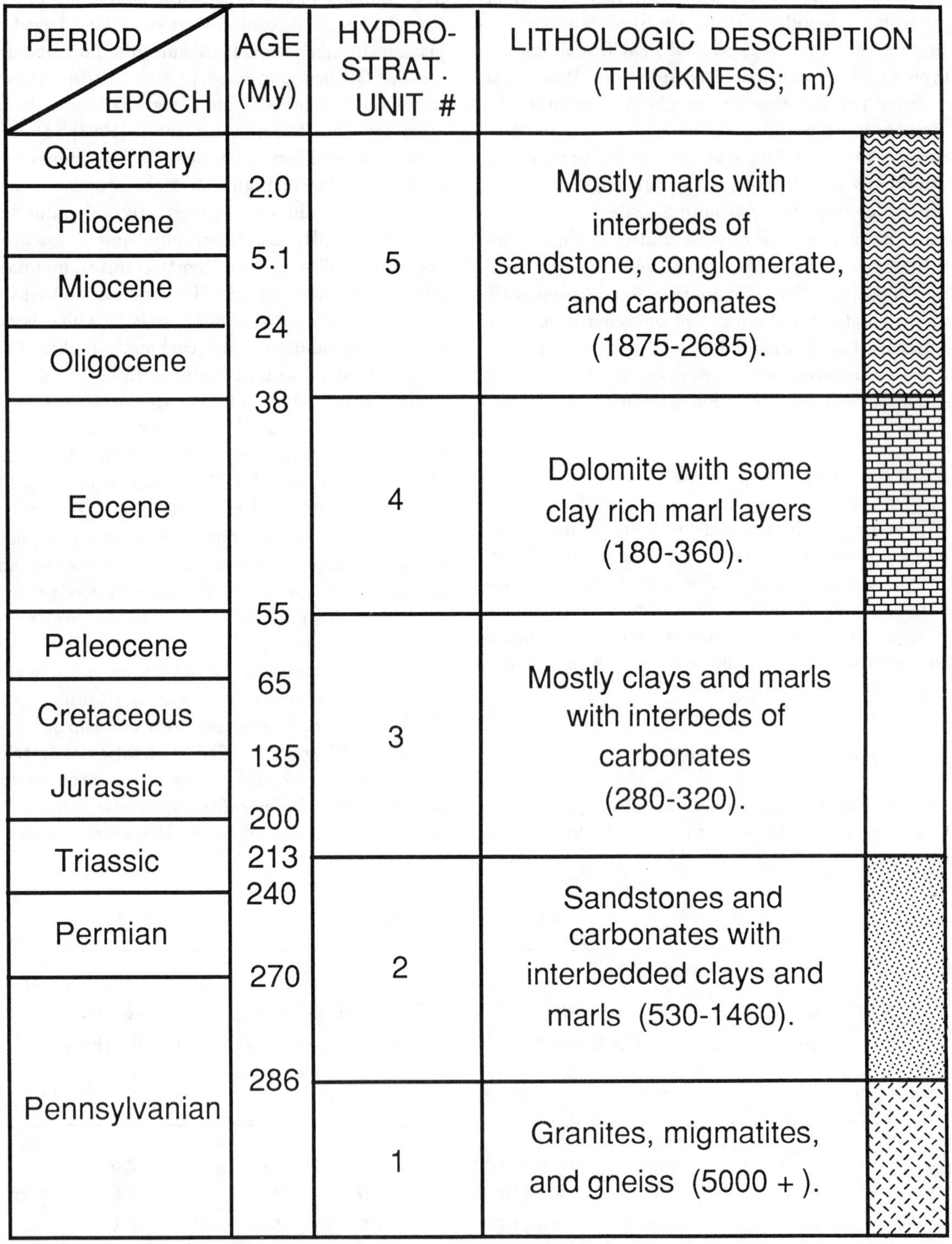

PERIOD / EPOCH	AGE (My)	HYDRO-STRAT. UNIT #	LITHOLOGIC DESCRIPTION (THICKNESS; m)	
Quaternary	2.0	5	Mostly marls with interbeds of sandstone, conglomerate, and carbonates (1875-2685).	
Pliocene	5.1			
Miocene	24			
Oligocene	38			
Eocene		4	Dolomite with some clay rich marl layers (180-360).	
	55			
Paleocene	65	3	Mostly clays and marls with interbeds of carbonates (280-320).	
Cretaceous	135			
Jurassic	200			
Triassic	213			
	240	2	Sandstones and carbonates with interbedded clays and marls (530-1460).	
Permian	270			
	286			
Pennsylvanian		1	Granites, migmatites, and gneiss (5000 +).	

Fig. 7. Generalized hydrostratigraphic column including lithologic description, age, and hydrostratigraphic unit numbers. The rock properties associated with each of the hydrostratigraphic units are given in Table 1.

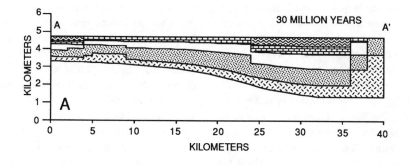

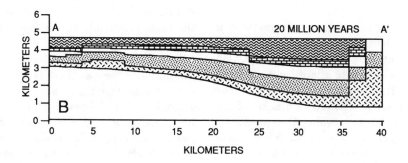

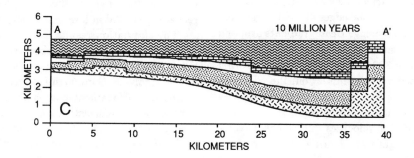

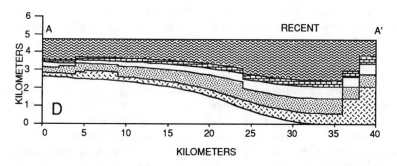

Fig. 8. Computed subsidence within the Rhine Graben, Pechelbronn Subbasin, along $A - A'$ over the last 30 million years (see Figure 1b for the location of section $A - A'$). Subsidence rates were specified for each nodal column with an average subsidence rate of $3 \times 10^{-5}\ myr^{-1}$. Rock properties and description of the hydrostratigraphic units are given in Table 1 and Figure 7.

observed fault block geometry is quite good (compare Fig. 1c with Fig. 8d). Faults were not represented as separate hydrostratigraphic units in our simulations.

The expansion of the finite element mesh during basin development is allowed by having the top row of elements grow to some specified thickness before a new row of elements is generated (Fig. 9). Sedimentation is thus allowed to keep pace with subsidence. Once a new row of elements is generated, the row thickness decreases through time due to sediment compaction. The addition of fluid mass and energy was taken into account in equations (1) to (3) for the top row of elements as they grew through time.

During the first 25 Myr, compaction and aquathermal pressuring dominate the hydrologic system (Fig. 10a and 10b). Hydraulic gradients due to sediment compaction are small because of the slow rate of sedimentation and permeable sediments. The largest head gradients occur near the land surface as fluids encounter a blanket of low permeability marls and clay sediments. Overpressuring occurred in the deepest part of the basin, but was nowhere greater than 10^{-3} m. During the later stage of basin development (Fig. 10c and 10d), subareal conditions allowed for the onset of a gravity-driven flow system. The largest head gradients occurred near the edges of the basin where prescribed heads were increasing. Head gradients dropped off quickly towards the center of the basin. Computed hydraulic heads varied over 300 m throughout the rift during the last 15 Myr. Based on this simulation, the effects of compaction-driven flow are completely overwhelmed by the onset of the gravity-driven flow system.

Computed groundwater velocities during basin evolution are presented in Figure 11. During the early phase of basin evolution (40 to 15 Myr ago), computed velocities were nowhere greater than 10^{-4} myr^{-1}. Near the top of the flow system, fluids expelled from compacting sediments moved directly upwards across the top boundary. Deeper within the basin fluids were focused into the basal aquifer and migrated to the edges of the basin before discharging. Between 15 and 0 Myr ago, groundwater flow patterns changed dramatically. Fluid-flow patterns reversed in direction as groundwater entered the rift along the edges of the basin and descended into the center of the basin before discharging. Computed velocities were highest near the uplifted escarpments of the rift where the water-table gradients were greatest. Predicted groundwater velocities are as high as 10^{-1} myr^{-1}.

The thermal evolution of the basin is presented in Figure 12. During the first 25 million years of basin formation, fluids expelled by compacting sediments had little effect on the temperature distribution as heat transfer was dominated by conduction. The effects of fluid flow can easily be seen when subsurface temperatures computed from the hydrothermal model are compared to temperatures calculated assuming conductive heat transfer alone. Temperatures based on conductive heat transfer are presented in Figure 13 for all four time steps. The effect of porosity on the bulk-thermal conductivity is also reflected in the temperature profiles in Figure 12a and 12b and Figure 13. The temperature contours tend to bunch up near the land surface due to a decrease in thermal conductivity with increasing porosity. Following marine regression, the onset of the gravity-driven groundwater flow system had a pronounced effect on heat transfer within the Rhine Graben. Fluids descending near the edges of the basin resulted in cooler temperatures than predicted by conductive heat transfer (compare Fig. 12c and 12d with Fig. 13c and 13d). Deviations from a conductive thermal field were as great as 60K in areas of groundwater recharge near the Graben framing faults. Upwelling groundwater in the center portion of the rift resulted in a broad region of gradually increasing high flow during the last 15 Myr (Fig. 14). Present-day computed heat flow is over 120 $mW\,m^{-2}$ in the center portion of the rift and s to less than 20 $mW\,m^{-2}$ at the fault escarpments. Near fault escarpments, convective-heat transfer due to groundwater recharge resulted in low heat flow during this time period.

The simulation results for present-day conditions are in general agreement with observed temperature data for the Rhine Graben. Some inconsistencies, however, could not be resolved. Comparison of present-day observed and computed surface-heat flow is shown in Figure 15. The most notable inconsistency is that the simulation model predicted lower heat flow (cooler temperatures) in the recharge areas than is observed in the field data. Another inconsistency is that the observed heat flow pattern is asymmetric, with highest heat flow associated with the Pechelbronn oil fields. The computed surface-heat flow, on the other hand, has a more symmetric pattern. There are several possible explanations for these inconsistencies. The incorporation of highly permeable fault zones within the center of the Graben may improve agreement between computed and observed temperature patterns. Doebl et al. [1974] and Otto and Tóth [1988], document areas of high heat flow near fault zones that are associated with geothermal springs within the rift. Conditions of high basal heat flow near the framing faults of the rift could help to explain this inconsistency as well. However, there is no evidence of recent magmatism along the Pechelbronn section to justify this boundary condition. It is also possible that along-axis fluid flow and heat transport may help to explain this discrepancy. Further sensitivity studies which incorporate permeable faults may help to resolve some of these discrepancies.

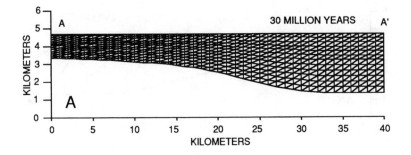

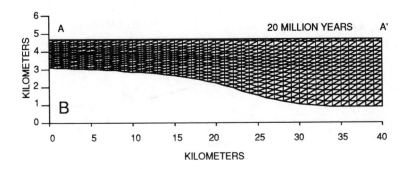

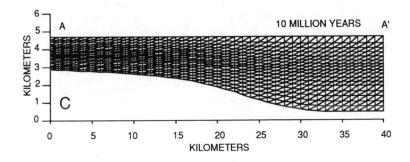

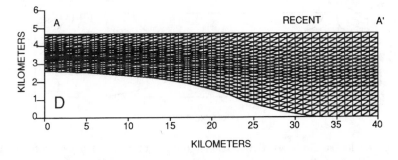

Fig. 9. Expansion of finite element mesh during basin evolution. The top row of elements were allowed to grow with subsidence to some specified thickness before a new row of elements were generated. Element areas decrease with depth to account for the effects of sediment compaction.

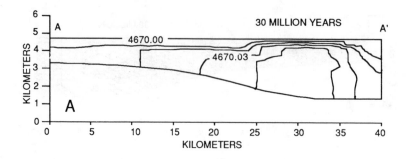

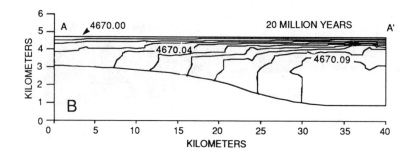

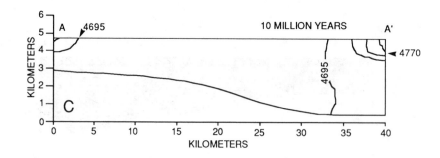

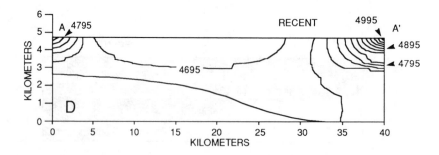

Fig. 10. Computed hydraulic head distribution (contour interval in m) within the Rhine Graben along $A - A'$ over the last 30 million years. In Figure 10a and 10b computed hydraulic gradients are due to sediment compaction and aquathermal pressuring. In Fig. 10c and 10d, computed head differences result from gravity-driven groundwater flow.

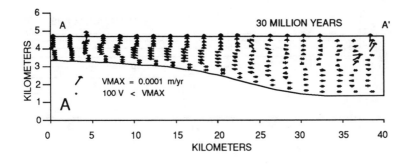

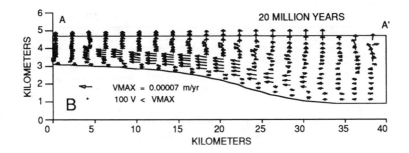

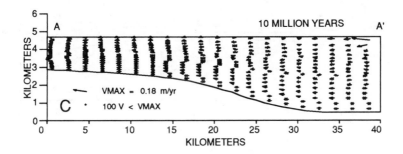

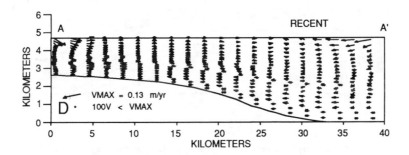

Fig. 11. Computed groundwater velocity distribution (m yr^{-1}) within Rhine Graben along $A - A'$ over the last 30 million years. During early stages of basin development (Fig. 11a and 11b), velocity vectors point mostly upwards and towards the edge of the basin due to groundwater flow driven by sediment compaction. During the last 15 My, groundwater flow directions are reversed due to the onset of a gravity-driven flow system (Fig. 11c and 11d).

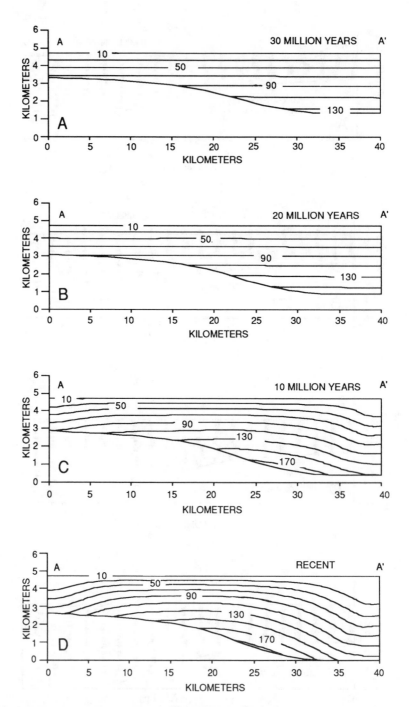

Fig. 12. Computed temperature distribution (°C) for Rhine Graben along $A - A'$ over the last 30 million years. During the early stages of basin formation, isotherms are flat as groundwater flow driven by sediment compaction was too small to disturb the conductive temperature field (Fig. 12a and 12b). Convective-heat transfer during the last 15 My resulting from gravity-driven flow caused significant deviations from a conductive thermal field. Relatively lower temperatures are found in groundwater recharge areas near fault escarpments while discharge areas within the center of the rift have higher temperatures (Fig. 12c and 12d).

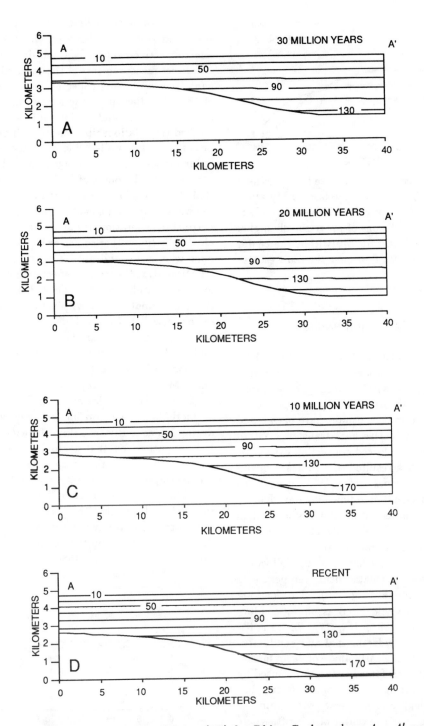

Fig. 13. Computed temperature distribution ($^{\circ}C$) for Rhine Graben along $A - A'$ over the last 30 million years assuming heat transfer by conduction alone. Temperature gradients are highest near the land surface due to increasing bulk-thermal conductivity and associated decreasing porosity with depth as computed by Equations (4) to (7).

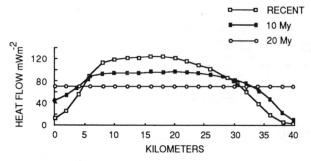

Fig 14. Temporal variations in computed surface-heat flow within the Rhine Graben over the last 20 My along the section $A - A'$. The onset of gravity-driven groundwater flow has resulted in conditions of progressively higher high heat flow during the last 15 My within the center of the Graben and lower heat flow near the fault escarpments.

Thermal Maturation of Rift Sediments

Results from the hydrothermal model of the Rhine Graben have important implications for the interpretation of geothermic data collected within the rift. For example, calculations of the thermal maturity of organic material in sediments during basin formation usually have been based on the assumption of conductive heat transfer. Within the Rhine Graben, however, forced convection caused by deep groundwater flow has resulted in significant deviations from a conductive thermal profile. While it has been recognized that hydrothermal fluid flow may have had an effect on sediment diagenesis within the Rhine Graben [Espitalie, 1984], no attempt has been made to quantify the impact of convective-heat transfer.

The oil generation window, defined here by TTI values between 15 and 160, was calculated using temperatures from the hydrothermal model (Fig. 16). The location of the oil window was also calculated assuming conductive temperatures for comparison (Fig 17). There is little difference between the hydrothermal and conductive maturation indices during the early phase of basin formation (first 25 Myr). However, forced convection caused by deep groundwater flow during the Miocene to Recent has resulted in regions of high and low heat flow within the rift. In discharge areas near the center of the rift, where heat flow is highest, the oil window occurs at a shallower depth than predicted by the conductive model (compare Fig. 16c and 16d to 17c and 17d). Conversely, in areas of groundwater recharge near the rift escarpment, where heat flow is lower, the oil window is deeper than would be predicted by a conductive thermal model. These findings

are consistent with Willett and Chapman [1987] who observed an association between vitrinite reflectance levels and convective-heat transfer within the Uinta Basin.

Direct measurements of the thermal maturation of organic rich sediments collected from the middle segment of the rift adjacent to the study area support the computed petroleum maturation indices. Doebl et al. [1974] studied the relationship between sediment diagenesis and heat flow in two wells adjacent to the study area. Calculations made using clay diagenesis and coalification data suggest that the onset of high heat flow began no earlier than 15 Myr ago in that part of the Graben. In one of the wells which had a geothermal gradient of 80 mKm^{-1}, petroleum generation began at a depth of 700 m [Doebl et al., 1974]. In another cooler well (40 mKm^{-1}), hydrocarbons were not found until a depth of 2600 m. Espitalie [1984] analyzed vitrinite reflectance data obtained from four wells within the Rhine Graben in order to reconstruct the geothermal history within the rift (see Fig. 5 for location of wells). He concluded that geothermal gradients could not have been constant through time. Using thermal modeling of organic maturation, he estimated that the onset of high heat flow probably began between 22 and 2 Myr ago (Fig. 18).

Both of the studies mentioned above document that an increase in heat flow has occurred within the Graben during Miocene-Pliocene time. Based on our calculations, it is hypothesized that tectonic uplift during the past 15 Myr, in conjunction with marine regression, lead to the development of a gravity-driven flow system that was responsible for the onset of high heat flow within the rift. It is unlikely that the observed increase in heat flow is due to temporal variations in basal heat flux as the lithosphere underlying the rift has not been appreciably thinned [Villemin et al., 1986].

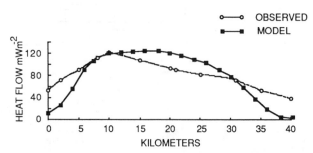

Fig 15. Comparison of present-day observed and computed surface-heat flow within the Rhine Graben along the section $A - A'$. The largest discrepancies between observed and computed heat flow occurs in the recharge areas near the fault escarpments.

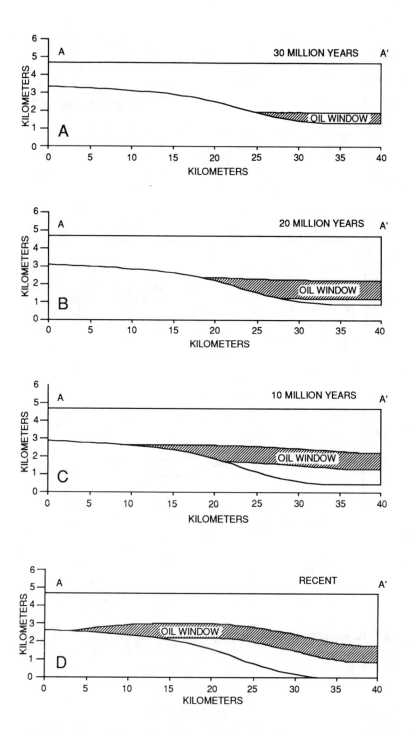

Fig. 16. Computed location of the oil window (TTI values between 15 and 160) along $A - A'$ over the last 30 million years using temperatures from the hydrothermal model.

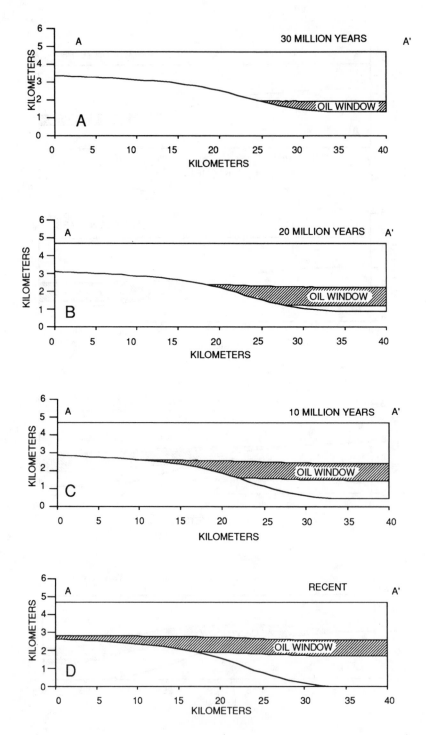

Fig. 17. Computed location of the oil window (TTI values between 15 and 160) along $A - A'$ over the last 30 million years using temperatures calculated assuming heat transfer by conduction alone.

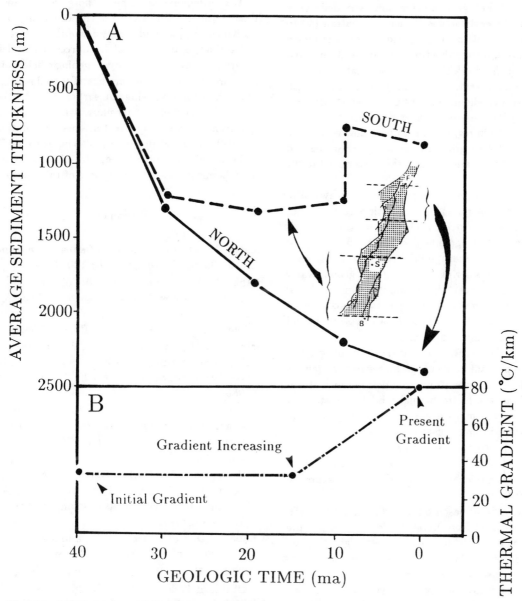

Fig. 18. (A) Average sediment thickness (in m) of rift sediments for the northern and southern depocenters [after Villemin et al., 1986]. Decrease in sediment thicknesses in the southern depocenter indicate uplift and erosion during Miocene-Pliocene times. (B) Observed increase in geothermal gradient within the middle segment of the Rhine Graben. Data is based on averaged paleogeothermal gradients estimated from vitrinite reflectance data of three boreholes adjacent to the study area [after Espitalie, 1984]. Tectonic uplift and marine regression during Miocene to Recent (Fig. 18a) may have induced a gravity-driven flow system responsible for the increase in thermal gradients (Fig. 18b).

Conclusions

1. Results from a paleohydrologic model of the Rhine Graben indicate that groundwater flow was driven by compaction in the early stages of basin formation (40 to 15 Myr before present) during a period of marine incursion. Heat transfer during this time period was dominated by conduction as fluid velocities were too small to disturb the thermal field. Marine regression in conjunction with uplift of the Graben shoulders during the last 15 Myr initiated a gravity-driven groundwater flow system. Forced convection resulted in regions of high heat flow within the center of the rift where groundwater was discharging and negative thermal anomalies in recharge areas near the rift margins.

2. Present-day temperatures computed from the hydrothermal model are in general agreement with observed data; however, some discrepancies could not be resolved. It is possible that better agreement between observed and computed present day thermal data could probably have been achieved by considering: (a) along-axis groundwater flow and energy transport, (b) spatial variations in basal heat flux, and most importantly, (c) better knowledge of the actual distribution and hydraulic nature of faults and of patterns of the temperature field.

3. The impact of convective-heat transfer on oil generation has had important implications for petroleum genesis within the Rhine Graben. Computed petroleum maturation indices from the hydrothermal model indicate that oil generation occurs much deeper in areas of groundwater recharge than in areas of groundwater discharge due to the impact of advective-heat transfer. These findings may also be applicable to other sedimentary basins which have undergone significant periods of subareal exposure and regional fluid flow.

4. Geothermic studies based on clay diagenesis and vitrinite reflectance data from boreholes located within the middle segment of the Rhine Graben are in good agreement with hydrothermal model results. These studies indicate that the onset of high heat flow occurred during Miocene-Pliocene times. Analysis of hydrocarbon occurrence within two boreholes indicates that the depth to oil generation is correlated with heat flow. Oil generation occurred at a depth of 2600 m in one well experiencing nominal geothermal gradients. Oil occurred at a much shallower depth (700 m) in another well that had higher heat flow.

5. Analysis of the hydrodynamic evolution within sedimentary basins using numerical analysis is a useful approach in assessing the thermal history of sedimentary basins. Probably the greatest shortcoming of the technique is the lack of availability of high quality geothermic and hydraulic data for full validation of the numerical results.

Acknowledgments. The authors would like to thank Jozséf Tóth and Claus Otto of the University of Alberta, and Klaus Trippler and Max Eder of the Federal Institute for Geoscience and Natural Resources, Federal Republic of Germany, for providing hydrogeologic field data for the study area. The authors are grateful to Jozséf Tóth and P. Y. Shen for reviewing an earlier draft of the paper. The work reported in this paper was supported in part by a grant from the National Science Foundation (EAR-8553019). Acknowledgement is also made to the Donors of the Petroleum Research Fund, administered by the American Chemical Society for support of this research.

References

Bentor, Y., On the evolution of subsurface brines in Israel, Chemical Geology, 4, 83-100, 1969.

Bethke, C. M., A numerical model of compaction-driven groundwater flow and heat transfer and its application to paleohydrology of intracratonic sedimentary basins, Jour. Geophys. Res., 90, no. B8, 6817-6828, 1985

Bethke, C. M., Hydrologic constraints on the genesis of the Upper Mississippi Valley mineral district from Illinois basin brines, Econ. Geol., 81, no. 2, 233-249, 1986.

Bredehoeft, J. D. and I. S. Papadopulos, Rates of vertical groundwater movement estimated from the earth's thermal profile, Water Resour. Res., 1, no. 2, 325-328, 1965.

Celal Sengor, A. M., Burke, K., and Dewey, J. F., Rifts at high angles to orogenic belts: Tests for their origin and the Upper Rhine Graben as an example, Am. Jour. Sci., 278, 24-40, 1978.

Chapman, D. S., Clement, M. D., and Mase, C. E., Thermal regime of the Escalante Desert, Utah, with an analysis of the Newcastle Geothermal System, Jour. Geophys. Res., 86, no. B12, 11735-11746, 1981.

Clauser, C., Conductive and convective heat flow in the northern Upper Rhinegraben- can they be separated?, (abs), International Union of Geodesy and Geophysics, XIX, General Assembly Proceedings, Vancouver, 1987.

Craig, H., Isotopic composition and origin of the Red Sea and Salton Sea geothermal brines, Science, 154, 1544-1548, 1966.

Craig, H., 1969, Geochemistry and origin of Red Sea brines, in Hot Brines and Recent Heavy Metal Deposits in the Red Sea, Edited by E. T. Degens and Ross, D.A., Springer-Verlag, 208-242, 1969.

Doebl, F., Heling, D., Homann, W., Karweil, J., Teichmuller, M., and Welte, D., Diagenesis of Tertiary clayey sediments and included dispersed organic matter in relationship to geothermics in the Upper Rhine Graben, in Approaches to Taphrogenesis, Edited by J. Illies, and Fuchs, K., Stuttgart, 192-207, 1974.

Edel, J. B., Fuchs, K., Gelbke, C., and Prodehl, C., Deep structure of the southern Rhinegraben area from seismic refraction investigations, Jour. Geophys., 41, 420-432, 1975.

Espitalie, J., Tentative reconstruction of geothermal paleogradients in some wells of the Rhine Graben, in Collection Colloques et Séminaires, 41, Edited by B. Durand, Institut Français du Pétrole, 147-166, 1984.

Garven, G., and Freeze, R. A., The role of regional fluid flow in the formation of ore deposition in sedimentary basins: A quantitative model, Nat. Hydrogeol. Conf., 2nd Ann. Meet., Winnepeg, Man., Proc., 60-67, 1982.

Garven, G. and Freeze, R. A., a, Theoretical analysis of the role of groundwater flow in the genesis of stratabound ore deposits : 1. Mathematical and numerical model, Am. Jour. Sci., 284, 1085-1124, 1984.

Garven, G. and R. A. Freeze, b, Theoretical analysis of the role of groundwater flow in the genesis of stratabound ore deposits : 2. Quantitative results, Am. Jour. Sci., 284, 1125-1174, 1984.

Groten, E., Gerstenecker, C., and Hein, G., Geodetic high- precision measurements in active tectonic areas; example: The Rhinegraben, Tectonophysics, 52, 157-165, 1979.

Haenel, R., Atlas of subsurface temperatures in the European Community, Commission of the European Communities, Hannover, 49 plates, 1979.

Huyakorn, P.S., and Pinder, G.F., Computational Methods in Subsurface Flow, New York, Academic, 473pp., 1983.

Illies, J.H., Graben tectonics as related to crust mantle interactions, in Graben Problems, International Upper Mantle Project, Scientific Report No. 27, in Edited by J. H. Illies and Mueller, S.T., Stuttgart, 4-27, 1970.

Illies, J.H., Two stages Rhinegraben rifting, in Tectonics and Geophysics of Continental Rifts, Edited by I. B. Ramberg and E. R. Neumann, Reidel Publishing Company, Dordrecht, Holland, 63-71, 1978.

Illies, J.H., and Greiner, G., Rhine Graben and the Alpine system, Geol. Soc. Am. Bull., 89, 770-782, 1978.

Issar, A., The paleohydrology of southern Israel and influence on the flushing of the Kurnub and Arad Groups (Lower Cretaceous and Jurassic), Jour. Hydrology, 44, 289-303, 1979.

Karrenberg, H. and Struckmeier, W., International hydrogeologic map of Europe, Int. Assoc. Hydrogeol. Comm. Geo. Map World., Bundesanstalt fur Geowissenschaften und Rohstoffe and UNESCO, Plate C4, 1970.

Lewis, T. J. and Beck, A. E., Analysis of heat flow data-detailed observations in many holes in a small area, Tectonophysics, 41, 41-59, 1977.

Lopatin, N. V., Temperature and geologic time as factors in coalification (in Russian), Akad. Nauk SSSR Izv. Ser. Geol., 3, 95-106, 1971.

Malzer, H. and Schlemmer, H., 1975, Geodetic measurements and recent crustal movements in the Southern Upper Rhinegraben, Tectonophysics, 29, 275-282, 1975.

Manheim, F. T., Red Sea geochemistry, in Initial Reports of the Deep-Sea Drilling Project, U.S. Govt. Printing Office, 23, 975-998, 1974.

Majorowicz, J. A., and Jessop, A. M., Regional heat flow in the Western Canadian Sedimentary Basin, Tectonophysics, 41, 209-238, 1981.

Morgan, P., Heat flow in rift zones, in Inter-Union Comm. on Geodyn. Ser., Am.Geophys. Union, 8, Edited by G. Palmason, 107-122, 1982.

Morgan, P., Harder, V., Swansberg, C. A., and Daggett, P. H., 1981, A groundwater convection model for Rio Grande rift geothermal resources, Geothermal Resour. Council Trans., 5, 193-196, 1981.

Otto, C., and Tóth, J., Hydrogeology of the Pechelbronn oil field, Rhine Graben, in Hydrogeology of Sedimentary Basins: Application to Exploration and Exploitation, Edited by B. Hitchon, S. Bachu, and C. M. Sauveplane, NWWA and Alberta Research Council, Edmonton, 105-106, 1988.

Person M. and Garven, G., Paleogroundwater flow and thermal evolution of the Rhine Graben, (abs), International Union of Geodesy and Geophysics, XIX, General Assembly Proceedings, Vancouver, 1, 78, 1987.

Roll, A., Versuch einer volumenbilanz des Oberrheintalgrabens und seiner schultern. Jahrb., Reihe A., 52, 82pp., 1979.

Schnabele, R., Hass, J. O., Hoffmann, C. R., Monographie Géologique du Champ Pétolifére de Pechelbronn, Mémoires du Service de la Carte Géophysique d' Alsace et de Lorraine, 7, 254pp., 1948.

Smith, L., and Chapman, D. S., On the thermal effects of groundwater flow. 1. Regional scale systems, Jour. Geophys. Res., 88, 593-608, 1983.

Stallman, R. W., Computation of groundwater velocity from temperature data, U.S. Geol. Surv. Water Supply Paper 1544-H, 36-46, 1963.

Tóth, J., A theory of groundwater motion in small drainage basins in central Alberta, Jour. Geophys. Res., 67, 4375-4387, 1962.

Tóth, J., Cross formational gravity-flow of groundwater: A mechanism of the transport accumulation of petroleum migration, in Am. Assoc. Pet. Geol., Studies in Geology, Problems in Petroleum Migration, 10, 121-167, 1980.

Villemin, T., Alvarez, F., and Angelier, J., The Rhinegraben: extension, subsidence, and uplift, Tectonophysics, 128, 47-59, 1986.

Waples, D., Time and temperature in petroleum formation: application of Lopatin's method to petroleum exploration, Am. Assoc. Pet. Geol. Bull., 64, no. 3, 916-926, 1980.

Willett, S. D., and Chapman, D. S., Temperatures, fluid flow and the thermal history of the Uinta Basin, in Migration of Hydrocarbons in Sedimentary Basins, edited by B. Doligez, Editions Technip, Paris, 533-552, 1987.

Zucca, J. J., The crustal structure of the southern Rhinegraben from reinterpretation of seismic refraction data, Jour. of Geophys., 55, 13-22, 1984.

CONDUCTIVE AND CONVECTIVE HEAT FLOW COMPONENTS IN THE RHEINGRABEN AND IMPLICATIONS FOR THE DEEP PERMEABILITY DISTRIBUTION

Christoph Clauser[1]

Institut für Angewandte Geophysik, Petrologie und Lagerstättenforschung,
Technische Universität Berlin, Ackerstr. 71-76, D-1000 Berlin 65,
Federal Republic of Germany

EXTENDED ABSTRACT

In order to assess quantitatively the amount of convection (free and forced) in the net heat flow density of the northern Rheingraben in southwest Germany, three methods were employed in Clauser (1988) that greately differ in their complexity and data-requirements: (1) an averaging technique based on temperature data from 164 shallow (<300 m) boreholes in the topmost quarternary aquifers of the Rhein valley, (2) a vertical Peclet-number analysis of temperature logs from 27 deep boreholes (≥1000 m), and (3) detailed 2D finite-difference-modeling of the fully coupled flow and heat transport equations across the valley and down to the graben's crystalline base. Fig. 1 shows the location of the study area and of the boreholes involved.

Apparent heat flow densities were determined for the 164 shallow boreholes from an average thermal conductivity and a mean temperature gradient. The latter was determined for each borehole over its entire length using the maximum temperature at the bottom of the borehole and an average surface temperature of 10.7 °C which was established from 25-year averages of soil temperatures in depths of 2 to 100 cm at four stations of the German weather service in the area. The average thermal conductivity of 2.5 ±0.5 W m^{-1} K^{-1} of the quarternary sediments which was used is based on selected in-situ measurements of thermal conductivity in boreholes and at outcrops in sand and gravel pits as well as on estimates based on porosity and mineral content.

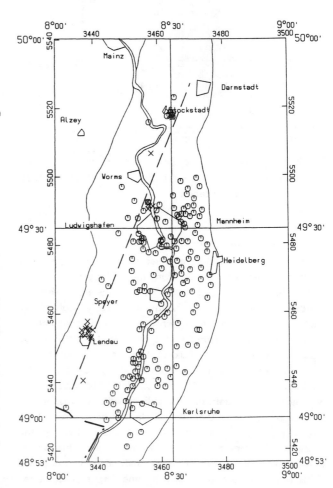

Fig. 1. Northern Rheingraben with distribution of shallow boreholes (circles) and deep boreholes (crosses), the broken line marks a potential SSW-NNE trending profile for the deep boreholes.

---

[1]Now at Gesellschaft für Strahlen- und Umweltforschung, Institut für Tieflagerung, Theodor-Heuss-Str. 4, D-3300 Braunschweig, Federal Republic of Germany

Copyright 1989 by
International Union of Geodesy and Geophysics
and American Geophysical Union.

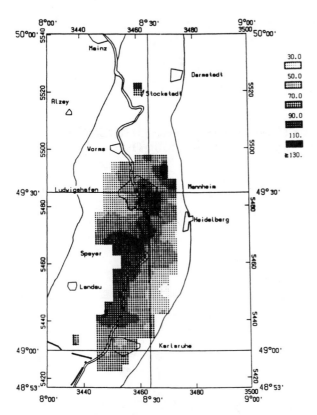

Fig. 2. Interpolated reprensentation of apparent surface heat flow density from 164 determinations in shallow boreholes of the northern Rheingraben.

Fig. 2 shows the resulting heat flow density pattern. It closely reflects the redistribution of heat within the upper two shallow quarternary aquifers from the recharge area at the valley flanks to the discharge zone with the river Rhein at the center of the valley. Although this clearly prohibits an analysis of single heat flow densities, the ensemble can be sensibly analyzed assuming that there is no convectively induced net influx or efflux of heat into or out of this aquifer system. This is not an unreasonable assumption since the base of the system is formed by thick tertiary sequences with high clay content providing them with a considerably smaller permeability than that of the aquifer. While these sequences crop out west of the river Rhein, leaving this boundary as impermeable as the base, the influx at the southern end of the system is approximately equal to the efflux at its northern end. For the remaining eastern valley flank the amount of flow across this boundary was estimated based on existing hydrological data, as was the amount of industrial water production from the aquifer. The amount of heat carried into and out of the system by these water flows is estimated to produce a systematic error in the determination of surface heat flow density of $+12$ mW m^{-2} and -26 mW m^{-2}

respectively. Thus a net error of -14 mW m^{-2} will be caused by treating the aquifer system as adiabatic. With this assumption, however, the apparent heat flow densities in Fig. 2 can be integrated over the entire area and division by the total area yields the average conductive basal heat flow density for this area; since a regular grid was used for the interpolation in Fig. 2 this procedure is equivalent of taking the mean value. The average basal heat flow density for this area was thus determined to of 82 mW m^{-2} with a standard deviation of ± 20 mW m^{-2}. It can be seen that the systematic error falls well into the range defined by the standard deviation, justifying the adiabatic simplification.

All of the 27 boreholes for the vertical Peclet-number analysis are aligned on a NNE-SSW trending profile in the area of the graben's western border faults. Applying Bredehoeft's & Papdopulos' (1965) half-space technique for a stratified medium the ratio Pe/L between the Peclet-number the length scale for vertical convection can be derived from the slope of a linear regression of the logarithmic heat flow density as a function of depth for each borehole: $\ln q(z) = \ln q_0 + z \cdot Pe/L$. q_0 is the surface heat flow density defined by the intercept at $z = 0$, and z as well as the direction of convection is taken positive downwards (see Fig. 3 for two examples). Since the two unknowns Pe and L cannot be determined from a single equation upper and lower bounds on L were used to derive upper and lower bounds on Pe as well as lower and upper bounds on the conductive basal heat flow density $q(L) = q_0 \cdot \exp(Pe)$. An obvious lower bound for L is always the borehole length; a sensible upper bound is the depth to the crystalline basement. The results of this analysis for those 25 boreholes that cluster in three definite areas (see Fig. 1) were again averaged yielding the final result displayed in Fig. 4: maximum and minimum basal heat flow densities with their respective Peclet-numbers along the 80 km profile. There is no obvious lateral trend in these results; the mean conductive basal heat flow density is 88 ± 18 mW m^{-2} and the average Peclet-number is 0.5 ± 0.2. This basal heat flow density agrees remarkably well with the one which was determined from the shallow boreholes in spite of the completely different data sets, methods used, and depths of the conductive base. It is also worth noting the physical meaning of a Peclet-number of 0.5, which is that two thirds of the total heat flow density are conducted to the surface leaving one third which is convected.

Finally, a detailed structural model of the Rheingraben was used to derive a starting model for 2D finite-difference simulations of the steady-state, fully coupled fluid flow and heat transport equations. Tabulated average values were taken for all physical parameters. Fig. 5 shows as an example the distribution of the hydraulic conductivity: prominent features are a relatively permeable sand- and limestone stratum dipping to the east, the eastern and

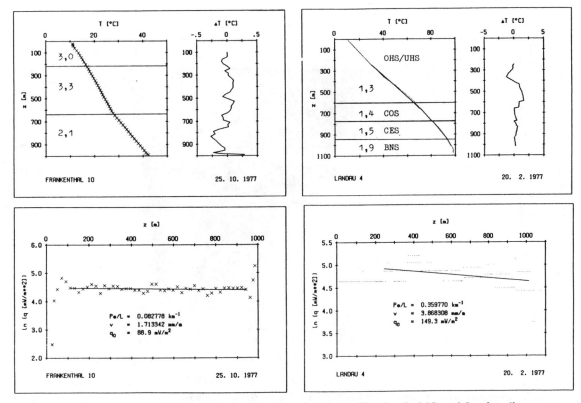

Fig. 3. Vertical Peclet-number analysis for two deep boreholes (Frankenthal 10 and Landau 4), crosses and dots mark measured temperatures, solid lines mark temperatures calculated from Peclet-number analysis results; ΔT refers to the difference between calculated and measured temperatures. Numbers in temperature profiles give average thermal conductivities in W m^{-1} K^{-1}, and abbreviations refer to specific sequences in the oligocene and miocene.

western fault zones, the thick, relatively low permeability tertiary sequence above it, as well as the crystalline flanks and basement of even lower permeability. Laterally, thermal and hydraulic no-flow boundaries were invoked. The base of the model is hydraulically a no-flow boundary as well, while three different basal heat flow densities of 60, 80, and 100 mW m^{-2} are specified. At the top of the model a constant temperature of 11 °C is used, as well as a constant hydraulic potential which varies laterally with the topography. Beginning with this starting model the permeability of the main aquifer is varied systematically over three orders of magnitude. This results in a total of 18 different models for all permeability distributions and basal heat flow densities. Fig. 6 shows a typical flow field displaying the two main features that are in common to all flow fields: (1) there is very little flow into the tertiary sequence (between km 50 and 75) which forms the base of the quarternary aquifer with the 164 shallow boreholes, and (2) there is predominantly vertical flow within the western fault zone (km 40 to 50). Thus the basic assumptions for the

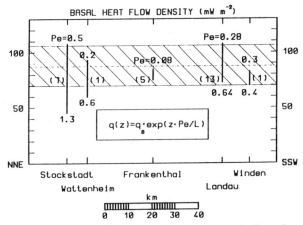

Fig. 4. Variation of Peclet-number and conductive basal heat flow density along the profile of deep boreholes in Fig. 1. Number in parentheses indicate number of determinations for each site. The broken line and hachured area indicate the average value and its standard deviation for all sites. The mean Peclet-number is 0.5 ± 0.2.

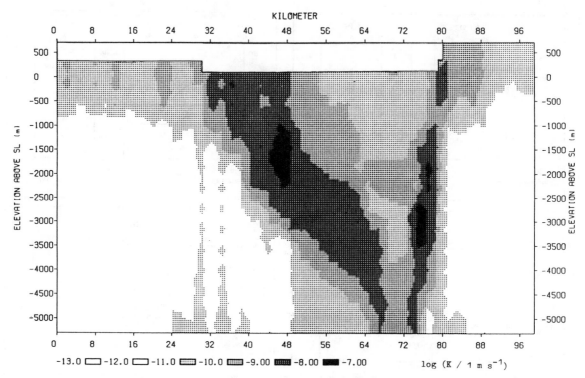

Fig. 5. Interpolated representation of hydraulic conductivity in model BGR9 on a 2D cross-section of the Rheingraben down to its crystalline base; solid line: earth's surface.

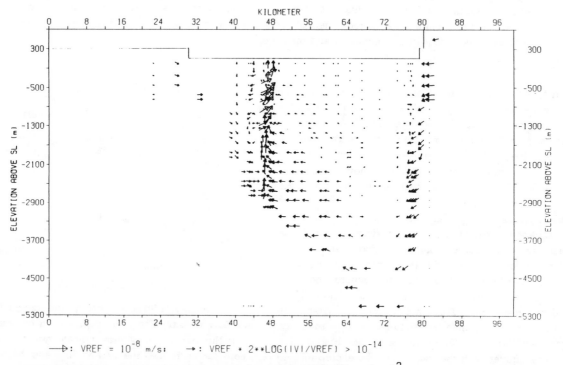

Fig. 6. Flow field results (Darcy-velocities) for model BGR9 (80 mW m^{-2} basal heat flow density); solid line: earth's surface.

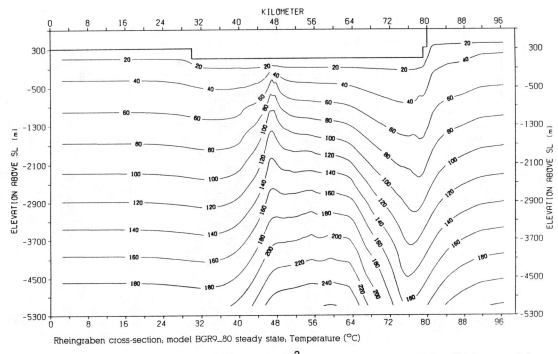

KILOMETER

Rheingraben cross-section; model BGR9_80 steady state; Temperature (°C)

Fig. 7. Temperature field for model BGR9 (80 mW m^{-2} basal heat flow density), solid line: earth's surface.

averaging technique as well as for the Peclet-number analysis are substantiated by the modeling results. Looking at the temperature field (Fig. 7) the well established east-west increase in the Rheingraben is reproduced as well as the elevated temperatures within the western fault zone, known from anomalies such as in Landau (Fig. 1) and Pechelbronn farther south in France. However, judging merely from the very sparse deep temperature information it is not possible to decide on the optimum model among the 18 models used. Therefore vertical Peclet- numbers were calculated for all 18 model results and vertical length-scales of L=1000 m and L= 2900 m. Fig. 8 shows the result plotted against the permeability of the main aquifer on a double-logarithmic scale for L=1000 m. It turns out that most of the models fail to lie within the Peclet-number interval of 0.5 ± 0.2 required by the results of the preceeding vertical Peclet-number analysis. Closer inspection of the results of those models that do reveals that these in turn fail to produce the required temperatures measured in boreholes within the western fault zone area (e.g. in Landau). It is therefore concluded that the adequate permeability can be constrained to within an order of magnitude between the distributions of model BGR10 (which satisfies the Peclet-number but not the temperature criterion) and model BGR9 (which reaches the required temperatures but displays Peclet-numbers slightly above the required range). The results of this partial calibration

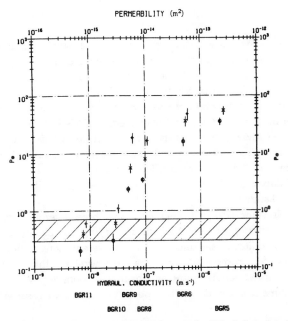

Fig. 8. Peclet-numbers for L=1000 m calculated for six different permeability distributions (represented by their mean value for the sand- and limestone aquifer and the fault zones: models BGR5 to BGR11) and three basal heat flow densities (squares: 60, crosses: 80, arrows: 100 mW m^{-2}) along the western fault zone area (km 40 to 50 in Figs. 5 to 7).

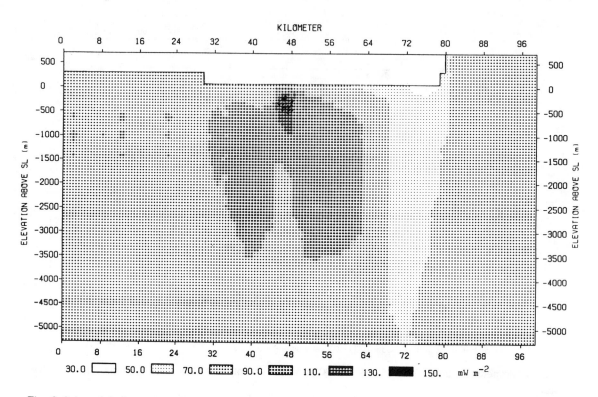

Fig. 9. Interpolated representation of apparent vertical heat flow density calculated from $\lambda \cdot \partial T/\partial z$ (λ: effective porous medium thermal conductivity) and temperature results in Fig. 7.

for L-2900 m are strikingly similar, but with considerably greater error bars for the Peclet-numbers due to the greater inhomogeneity of the flow field over this depth interval. Finally, Fig. 9 displays the distribution of a vertical heat flow density as derived from the temperature field in Fig. 7: above a basal conductive heat flow density of 70-90 mW m^{-2} the well known east-west increase in surface heat flow density can be recognized, as well as the elevated values of 110-150 mW m^{-2} in the western fault zone area and in the depth interval in which they were experimentally confirmed, e.g. in Landau.

In summary: joint interpretation of three different data sets employing different methods of varying complexity allows a quantitative separation of conductive and convective heat transfer components for the Rheingraben that satisfies observed temperatures and heat flow densities. While the numerical model-results support the flow-field assumptions for the averaging technique and the vertical Peclet-number analysis, the Peclet-number results can be used for a partial calibration of the hydrothermal model. Thus the uncertainty in the permeabilty of the aquifer can be reduced from the initial three orders of magnitude to one. This illustrates a potentially useful thermal method for constraining the permeability of deep strata.

References

Bredehoeft, J.D., and Papadopulos, I.S., 1965. Rates of Vertical Groundwater Movement Estimated from the Earth's Thermal Profile. Water Resour. Res., 1(2): 325-328.

Clauser, C. 1988. Untersuchungen zur Trennung der konduktiven und konvektiven Anteile im Wärmetransport in einem Sedimentbecken am Beispiel des Oberrheintalgrabens. Ph.D. thesis, Tech. Univ. Berlin, Fortschritt-Berichte VDI, Reihe 19, No. 28, VDI Verlag, Düsseldorf (Germany).

NUMERICAL THERMOHYDRAULIC MODELING OF DEEP GROUNDWATER CIRCULATION IN CRYSTALLINE BASEMENT: AN EXAMPLE OF CALIBRATION

Jean-Claude Griesser

Basler & Hofmann AG, Forchstr. 395, CH-8029 Zurich/Switzerland

Ladislaus Rybach

Institute of Geophysics, ETH-Hoenggerberg, CH-8093 Zurich/Switzerland

Abstract. A 2D integrated finite difference model of coupled fluid flow and heat transfer is presented for an area in northern Switzerland where crystalline basement is overlain by a Mesozoic/ Tertiary sedimentary sequence. In this region the geothermal field is characterized by a strongly positive heat flow density anomaly (> 150 mW m^{-2}). The anomaly is centered above a recently discovered Permocarboniferous trough. Several possibilities to explain the anomaly (thermal disturbance in the mantle, cooling shallow intrusion, locally strong uplift/erosion, local contrasts in petrophysical properties) can be ruled on in the basis of model calculations. Ascending deep groundwater is favored as the mechanism creating the observed anomaly. Deep groundwater circulation was investigated in detail, especially to clarify the hydraulic role of the Permocarboniferous trough, by coupled thermo-hydraulic modeling.

The model was calibrated to reproduce the data set of field measurements available to date. The following parameters were used: i) hydraulic potential of aquifers, ii) discharge rate and temperature of thermal springs, iii) surface heat flow density distribution, iv) subsurface temperature field (as constructed from drillhole data). The model so calibrated enables to delimit the depth range of deep groundwater circulation ($\sim$ 8 km), and the derivation of mean permeabilites and Darcy velocities for various parts of the crystalline basement (on the average, 10^{-15} m^2 and 10 mm/y, respectively, for the area studied).

Introduction

Coupled heat and fluid transfer modeling in fractured crystalline rock is of growing importance in fields like radioactive waste disposal or hot dry rock geothermal energy. The numerical models must be carefully calibrated in order to obtain realistic results. For the calibration a broad data base with different and independent sets of field measurements should be available.

An example of 2-D integrated finite difference modeling is presented in this study for an area in northern Switzerland. In this region, extensive drilling and seismic studies have been carried out within the framework of a high-level radioactive waste disposal feasibility study (Thury and Diebold, 1987). The same area, which is characterized by numerous thermal spring occurrences and by other manifestations of thermalism, was the target of detailed prospecting for geothermal resources (Gorhan and Griesser, 1988). From all these recent investigations a broad data base is available for calibration.

The Area Under Study

Geology

The study area is located in the central part of northern Switzerland, about 20 km northwest of Zurich in the region of the lower Aare valley (see Figure 1), and is characterised by the confluence of the three rivers Aare, Reuss and Limmat. Geologically it belongs to the eastern end of the Jura which is one of the three main tectonic units of Switzerland. The area is bordered in the north by the Hercynian Black Forest massif and in the south by the Tertiary Molasse basin.

The Black Forest massif, outcroping in the study area immediately north of the Rhein river, dips with an angle of 3-5° to form the crystalline basement below the Paleozoic and Mesozoic sediments of northern Switzerland. An additional element is an east-west trending Permocarboniferous trough (cf. Figure 1). This coal-bearing trough with a width of 10-12 km and a depth of probably up to 7 km was recently discovered and delimited by drilling and reflection seismic profiling (Laubscher, 1987).

The Jura consists of the two sub-units Folded

Copyright 1989 by
International Union of Geodesy and Geophysics
and American Geophysical Union.

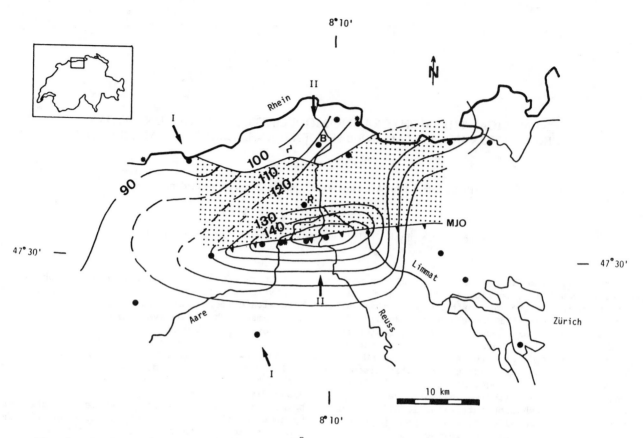

Fig. 1.   Surface HFD map (isolines in mW m^{-2}) of the area investigated. Black dots:
Drillholes with temperature and thermal conductivity measurements. Stars: Thermal water
occurrences. Dark area: Permocarboniferous trough. MJO: Main Jura Overthrust. B and R:
Drillholes Böttstein and Riniken (see text). Arrows I: Profile trace of Figure 2, arrows
II: Profile trace used in numerical modeling.

Jura in the south and Tabular Jura in the north,
formed by Mesozoic limestones, dolomites, marls and
evaporites. The two units are separated by the Main
Jura Overthrust, an east-west striking, southward
dipping, listric fault zone along which the Folded
Jura is thrusted over the Tabular Jura. Figure 2
shows a northeast-southwest profile through the
western part of the area under study. The profile
is based on the interpretation of the above-mentioned seismic measurements (Diebold, 1986). On one
hand the profile shows the complex structural style
of the Permocarboniferous trough, on the other the
close relationship of the southern border of the
trough with the Main Jura Overthrust. The sedimentary fill of the trough consists of consolidated
clastic rocks from Upper Carboniferous (290 m.y.)
to Lower Permian (250 m.y.) age (Matter, 1987).

Hydrogeology

Due to its low elevation the area under study
forms the drainage zone of river systems originating in the Alps and in the Black Forest massif.

Also for deep groundwater flow systems the lower
Aare valley and the Rhein valley represent a regional discharge area. The most important regional
aquifers are the karstified limestones and dolomites of the Upper Muschelkalk (Upper Triassic) and
the fractured top part of the crystalline basement.
Based on drillhole observations it seems that the
crystalline basement is more intensively fractured
north of the Permocarboniferous trough than south
of it. Beneath the trough, seismic interpretation
reveals a certain degree of fracturing (Laubscher,
1987). The permeable Upper Muschelkalk outcrops
along the Main Jura Overthrust zone and yields the
thermal water of the well-known spas Baden and
Schinznach with temperatures up to 47°C. Geothermal
investigations carried out along this zone (NEFF,
1984) indicate that the thermal water discharging
along the Main Jura Overthrust zone amounts in
total to about 0.02 kg/s per m length and that it
is a mixture of water from the Upper Muschelkalk
aquifer with recharge areas in the Jura or (further
south) in the Alps and of water from the crystalline basement, recharging probably in the Alps. The

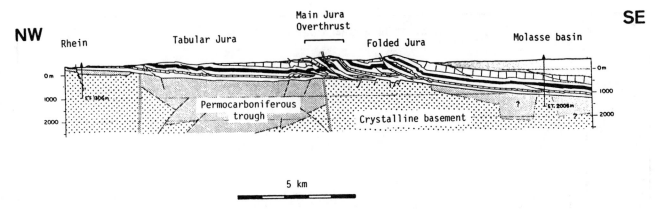

Fig. 2. Geologic section with the main tectonic units of the area (after Diebold, 1986).

age of this latter, deep thermal water component is more than 20'000 years (Gorhan and Griesser, 1988).

Geochemical studies (NAGRA, 1985; Schmassmann, 1987) show that in the Pre-Triassic formations (Permocarboniferous trough, crystalline basement) at least three deep groundwater types can be identified: 1) The water north of the trough in the fractured top of the crystalline basement is characterized by relatively low salinity ($\sim$ 1 g/l). It originates from the southern parts of the Black Forest massif and flows below the eastern Tabular Jura in northwestern direction to discharge in the Rhein valley; 2) in the trough itself highly saline water was found (> 50 g/l). 3) In the crystalline basement south of the trough the water originates from the Alps and is mineralised up to 10 g/l.

Geothermics

The geothermal regime in northern Switzerland is characterised by a strong positive temperature gradient and heat-flow density (HFD) anomaly (Figure 1). The HFD map of Figure 1 is based on temperature measurements in the boreholes indicated, some of them reaching down to depths of 2 km. The anomaly consists of two components with different wavelength: a broader component culminating with about 130 mW m^{-2} near the confluence of the three rivers Aare, Reuss and Limmat (cf. Figure 1), and a second, narrow component associated directly with the southern margin of the Permocarboniferous trough at the Main Jura Overthrust. In this zone HFD values up to 185 mW m^{-2} and temperature gradients up to 100 mK m^{-1} have been found (Griesser, 1985). The local thermal anomaly is caused by water ascending along the Main Jura Overthrust.

The amount, circulation depth and temperature of the water ascending along the overthrust is insufficient to create the entire anomaly displayed in Figure 1 (NEFF, 1984). Therefore, further explanations must be sought for the first HFD component. The discussion of interpretative possibilities to explain the observed long-wavelength anomaly in the

light of all available data is one of the aims of this paper.

Interpretation of the HFD Anomaly

The most prominent geothermal signature of northern Switzerland manifests pronounced, upward oriented heat transfer (in addition to the undisturbed background HFD which amounts to about 90 mW m^{-2}, cf. Bodmer and Rybach, 1984) in the study area. In principle, the long-wavelength surplus heat flux ($\sim$ 50 mW m^{-2}) could originate from conductive or convective heat transfer or both.

A pronounced positive surface HFD anomaly can have several explanations: a) strong local heat source in the mantle, b) cooling of a shallow intrusion, c) pronounced contrasts in petrophysical properties like thermal conductivity or radioactive heat production, d) rapid uplift and erosion and e) ascending deep groundwater. Processes b) and d) are clearly of transient nature whereas a), c) and e) could be related to steady-state phenomena. In the following, each of these interpretative possibilities will be judged on the basis of model calculations.

a) A thermal disturbance could exists in the mantle near the crustmantle boundary below the area in question with observable effects at the earth's surface. However, the limited lateral extent of the HFD anomaly (half width > 20 km) rules out this explanation: the crustal thickness in the area in question amounts to 25-30 km (Deichmann and Rybach, in press.).

b) A cooling intrusion in the upper crust of appropriate size (to match the surface HFD pattern) could give rise to the observed surface HFD distribution. Model calculations have been performed to match 2-D HFD anomaly profiles with a corresponding body (rectangular parallelepiped with horizontal dimensions: 11 x 16 km, thickness: 10 km, depth of top: 10 km), based on the procedure of Simmons (1967b). The results (Figure 3) reveal that the intrusion should be younger than 1.0 m.y. to explain

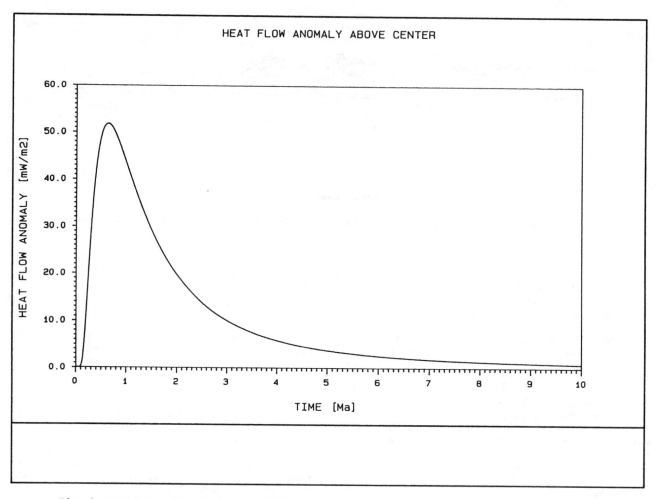

Fig. 3. Surface HFD anomaly created by an intrusive body above its center. Temperature of intrusion: 1200°C. For details see text.

the HFD anomaly. There are no signs of such a young magmatic activity known in the area (Rybach et al., 1987). This upper age limit would be further shortened by convective effects (additional cooling by circulating fluids).

c) The sedimentary trough of Permocarboniferous rocks, embedded in the crystalline basement of northern Switzerland, could increase the surface HFD if the sedimentary fill has high thermal conductivity or heat production relative to its surroundings. Figure 4 shows results of 2-D model calculation for a rectangular trough with the approximate dimensions given in Figure 2. The calculation is based on the "vertical sheet" formula of Simmons (1967a). Should the HFD anomaly be explained by anomalously high heat production, the Permocarboniferous sediments would have to possess four times the observed natural radioelement abundances (Rybach et al., 1987). The situation is similar with respect to a possible thermal conductivity contrast: the HFD anomaly can only be explained

with thermal conductivities of the Permocarboniferous sediments which are more than twice as high as observed (Rybach et al., 1987).

Anomalously high heat production of the basement below the trough could also create a surface HFD anomaly. Model calculations with anomalous bodies of different shapes reveal that a heat production contrast of > 15 μW m^{-3} would be needed to create the observed anomaly. Theres is no petrological evidence for such lithologies in the basement rocks of the area in question.

d) A locally strong uplift/erosion could also give rise to a positive HFD anomaly. The key parameter is the velocity of the process; it is normally assumed that the velocity of uplift equals that of erosion. Calculations for a simple 1-D situation, following the procedure described in Haenel and Zoth (1973), reveal that the uplift/erosion rate must be in the order of 10 mm/y over the last several 10'000 years to explain the HFD anomaly. Again, there are no signs (eg. by evidences in

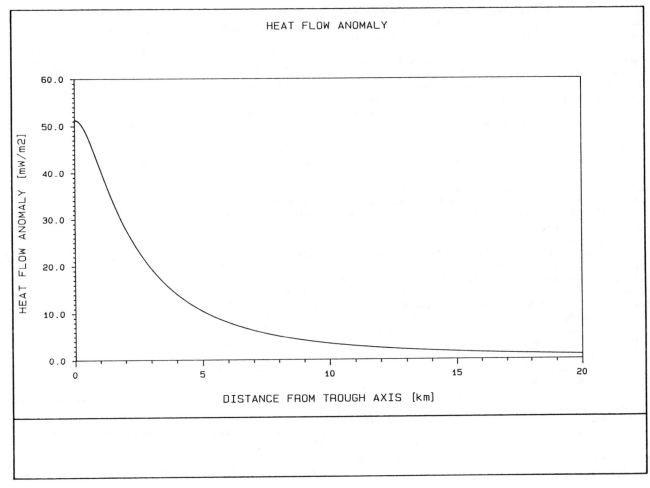

Fig. 4. Surface HFD anomaly profile across an infinitely long sedimentary trough with elevated heat production. Distance on horizontal axis is measured from center of trough. For details see text.

Quaternary geology) for such massive local vertical crustal movements in the area.

Since the possibilities a) to d) to explain the observed HFD anomaly can be ruled out, the only remaining option of HFD interpretation is the existence of groundwater ascending from greater depths in the area under study. In fact, there are several lines of evidence which point towards groundwater circulation with a significant vertical component:
- naturally issuing thermal water, especially along the Main Jura Overthrust zone (e.g. the spas Baden and Schinznach)
- a general decrease of geothermal gradient with depth, irrespective of lithology and tectonic setting (Rybach et al., 1987)
- a general increase of hydraulic potential with depth (as observed in several deep drillholes).

In the following, hydraulically plausible and geothermally significant deep groundwater circulation systems will be addressed in the realm of the Permocarboniferous trough of northern Switzerland. The systems should be modelled quantitatively in order to characterize them in terms of flow direction and flow rate and to delimit their depth range. Special emphasis shall be given to hydraulic effects (draining effect of the trough boundaries) as well as on the associated effects on the temperature field.

Numerical Modeling of Deep Groundwater Circulation

The previous section has shown that the favoured mechanism to explain the thermal anomaly in the region of the lower Aare valley is upflowing deep groundwater. The circulation system could involve the crystalline basement below the Permocarboniferous trough as well as the trough itself. In order to obtain a better insight into the hydrogeological conditions which would be necessary to

create such an anomaly, numerical two-dimensional, thermohydraulic simulations have been carried out. The following questions were considered in particular:
- what is the influence of the flow field, imposed by the boundary conditions and by the specific permeability distribution, on the subsurface temperature field and on the surface HFD distribution;
- which permeabilities and flow rates are necessary to generate the observed thermal effects.

Modeling Tool

The numerical simulations were carried out with the program PT (Pressure-Temperature) developed at the Lawrence Berkeley Laboratory, University of California (Bodvarsson, 1982). It can be used for one-, two- or three-dimensional coupled thermohydraulic numerical simulations of saturated, heterogenous, anisotropic and transient systems. The program is based on an Integrated Finite Difference algorithm for simultaneous determination of the pressure and temperature field. The P and T fields are coupled by pressure- and temperature-dependent rock and water properties. Program PT simulates steady-state conditions as the result of a transient problem after a very long simulation time. 2-D situations are modelled by slices (pseudo-section) with a nominal extension of 1 m in the third dimension.

The program PT is based on the following assumptions:
- fractured rocks can be approximated by equivalent porous media;
- the flow field is governed by Darcy's law;
- rock and pore fluid are in thermal equilibrium at all times.

Model Concept

Griesser (1985) has investigated, by means of two-dimensional northsouth trending sections (with a length of 17 km, a depth of 4 km and 1 m in the third direction; the trace of this profile is indicated in Figure 1), the influence of different flow directions in the crystalline basement in the model section on the subsurface temperature and pressure field. The presence of the Permocarboniferous trough with low-permeablity rocks was taken into account. It was only possible to match the observed temperature field if the water in the crystalline basement originates from recharges in the Alps; this calls for a general flow direction from south to north. The calculations have show that a HFD as high as 120 mW m^{-2} had to be assumed at a depth of 4 km in the center of the Permocarboniferous trough to match surface observations.

In order to shed light on the origin of such a high HFD at 3-4 km depth the model profile used by Griesser (1985) was extended to a depth of 8 km for this study. Figure 5 shows the geometry of this model. The calculations were based on the following conceptual assumptions: Below the trough or in its

TABLE 1. Hydrostratigraphic units of the model and their properties

Unit	Permeability (m^2)	Porosity (%)
Mesozoic aquitards	$1*10^{-17}$	5
Mesozoic aquifer (Upper Muschelkalk)	$5*10^{-13}$	10
Permocarboniferous trough	$1*10^{-17}$	3
Boundary zone of trough and upper crystalline basement north of the trough	$5*10^{-15}$	3
Upper crystalline basement south of trough	$5*10^{-16}$	1
Crystalline basement below the trough	$3*10^{-16}$	1
"Impermeable" deep crystalline basment	$1*10^{-18}$	1
Main Jura Overthrust zone	$5*10^{-14}$	5

deeper parts water is ascending from a depth of about 6-7 km along fracture zones. At 3-4 km depth, the water is mainly drained by the borders of the trough due to their higher permeability relative to the sedimentary fill.

This hydrogeologic model, assuming an enhanced permeability of the basement rocks below the trough, is based on the following evidences: noble gas contents and geochemical reservoir temperatures (> 80°C) of the deep component of the thermal water in Baden and Schinznach indicate a circulation depth of several kilometers (NEFF, 1984). Seismic reflection measurements (Laubscher, 1987) reveal that several major faults intersect within or in the vicinity of the Permocarboniferous trough in the area of the Lower Aare Valley. The deepest part of the trough is located in the same area which seems to be a result of subsidence along a deep-reaching fault system.

The depth of 6-7 km for the origin of ascending water was taken because seismic reflection measurements have identified fracture zones extending at least down to this depth. In the "Upper Muschelkalk" and the "top of the crystalline basement" aquifers the general flow direction was taken from south to north, in agreement with the regional pressure gradient direction.

The geometrical model (Figure 5) consists of 240 elements and is subdivided into 8 hydrostratigraphic units with contrasting permeabilities. For

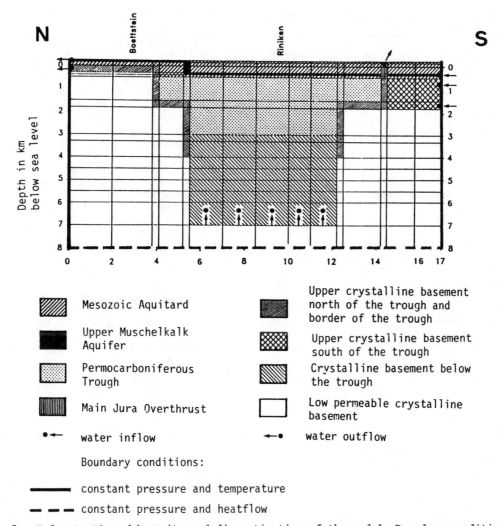

Fig. 5. Hydrostratigraphic units and discretisation of the model. Boundary conditions are also shown. For numerical values of hydraulic conductivity and porosity see Table 1.

the modeling a continuum representation was adopted for the hydrostratigraphic units. This assumption seems to be valid, in view of the high fracture density of the fractured zones and of the aquifers in question (NEFF 1984, NAGRA, 1985). The dimensions of the individual elements range from 100 to 2000 m. Table 1 shows numerical values used for the most relevant parameters of the hydrostratigraphic units. They were compiled from field and laboratory data (NEFF 1984, NAGRA 1985). During calibration of the model the permeabilities of some of the hydrostratigraphic units had to be adjusted slightly. In Table 1 the final values are given.

Boundary Conditions

The model used represents only a segment of the regional deep groundwater flow system between the Alps in the south and the Rhein river in the north. The hydraulic boundary conditions must therefore be chosen carefully in order to obtain a satisfactory fit to the observed hydraulic data (head distribution in the Upper Muschelkalk aquifer and in the crystalline basement, discharge rates of thermal water at the Main Jura Overthrust). For the northern and southern boundaries hydrostatic pressure conditions were chosen (see Figure 5 for the arrangement of the different boundary conditions applied). The model allows for water inflow at the southern model boundary into the Upper Muschelkalk and into the upper part of the basement as well as into the basement below the Permocarboniferous trough (see arrows in Figure 5). Since the origin of the water below the trough is not (yet) known a possible lateral inflow from the east or from the west can also be considered. Outflow of water is

TABLE 2. Comparison of measured and calculated
formation pressures

Drillhole	Depth (m)	Formation pressure (MPa) measured	calculated
Böttstein	400	4.2	3.9
	600	6.1	5.8
	800	8.0	7.5
	1350	14.8	14.0
Riniken	600	5.8[1]	5.9

[1] Calculated from a head measurement

only possible at the northern model boundary and
at the Main Jura Overthrust zone; all other boundary elements have "no flow" conditions.

The following thermal boundary conditions were
chosen: a constant HFD of 70 mW m^{-2} along the
bottom of the whole profile, and for the lateral
boundaries the temperature gradient was taken
according to the bottom (heat flow) and top (surface temperature) boundary conditions and to the
thermal conductivities of the different hydrostratigraphic units. The upper boundary corresponds to
an altitude of 250 m above sea level to which a
constant temperature of 13°C was assigned.

Calibration of the Model

In general, a model is successfully calibrated
if it is able to reproduce all observed field data.
The degree of calibration strongly depends on the
diversity and quality of the available data set.
For the proposed model the following field data
were available (Griesser, 1985): a) regional estimates of the hydraulic head gradients of the Upper
Muschelkalk aquifer and of the fractured top of the
crystalline basement, on the basis of water-level
measurements in relatively shallow drillholes: b)
detailed formation pressure measurements in the two
deep drillholes Böttstein and Riniken (for locations see Figure 1); c) temperatures and discharge
rates of the thermal water issuing along the Main
Jura Overthrust zone; d) surface HFD pattern (cf.
Figure 1); e) subsurface temperature field down to
a depth of 2 km as constructed from drillhole data;
f) permeability measurements in shallow drillholes
by pumping tests. High-quality perameability measurements (also for the aquitards) exist for the
drillholes Böttstein and Riniken. In Böttstein,
more than 100 permeability measurements were made
over a total length of 1200 m in the basement; g)
thermal conductivities for all relevant formations
have been measured on cores from all drillholes
indicated on Figure 1.

The pressure field is known only by a few but
high-quality values. Considerably more information
is available about the temperature field. No mea-

surements exist from depths greater than about
2 km.

The calibration itself was established by chan-
ging the pressure boundary conditions of the model
and by adjusting of the permeabilities of sensitive
hydrostratigraphic units. The thermal conductivities were not directly used for the calibration.
Tables 2 and 3 show a comparison of calculated and
measured formation pressures in the drillholeσ
Böttstein and Riniken as well as temperatures and
discharge rates of thermal water along the Main
Jura Overthrust.

Figure 6 shows the temperature field calculated
by the model. The agreement of calculated and observed temperatures down to a depth of 2 km is
reasonably good as can be seen from the drillhole
data of Riniken and Böttstein. The main directions
of fluid flow from the zone below the trough is
indicated schematically by arrows.

Results

The temperature field (Figure 6) is characterized by elevated temperatures at the Main Jura Overthrust zone, resulting from the upflow of thermal
water. This temperature anomaly is only of local
extent. In the region of the Permocarboniferous
trough a distinct upwelling of the isotherms is
evident.

In the northern part of the model the border
zone of the trough directs the upflowing water to
the permeable top part of the crystalline basement
for discharge into the Rhein river; in the south
the ascending water is led, together with crystalline water from the Alps and from the Upper
Muschelkalk aquifer, towards the Main Jura Overthrust zone.

The total amount of ascending water responsible
for this feature (crossing the whole width of about
6 km of the lower trough boundary) is 6*10^{-4} kg/s
per m in lateral direction. This value is the
result of the calibration process, and depends
mainly on the pressure applied at the water inflow
nodes and on the permeability of certain parts of
the basement. Because the general regional trend
of the hydraulic pressure gradient is towards the
north in the study area most of the ascending
water is drained through the northern boundary of
the trough. A smaller flow of water is directed
towards the southern boundary of the trough and

TABLE 3. Comparison of calculated and observed
thermal water discharges at the Main Jura
Overthrust zone

	Mean discharge rate (kg/m,s)	Mean temperature (°C)
Observed	0.007 - 0.06	39 - 47
Calculated	0.019	41.2

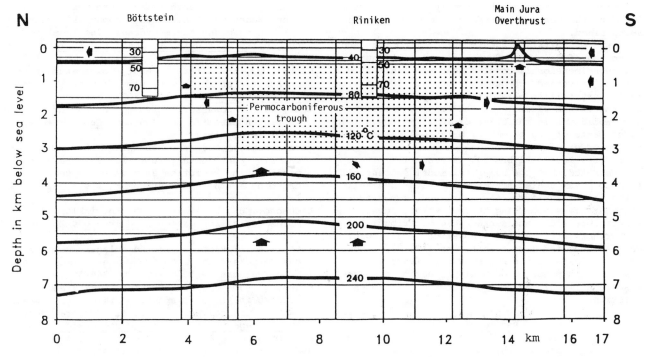

Fig. 6. Calculated temperature field in the realm of the Permocarboniferous trough. Dominant directions of the flow field are indicated schematically by arrows. Measured temperatures (drillholes Böttstein and Riniken) are shown for comparison.

discharges at the Main Jura Overthrust zone. The HFD calculated from the temperature field and from drillhole thermal conductivity data amounts to more than 120 mW m^{-2} at Riniken and to about 110 mW m^{-2} at Böttstein; at the northern model boudary the HFD is still high at 100 mW m^{-2}. Towards the southern boundary the HFD decreases (except for a local maximum at the Main Jura Overthrust) to values around 95 mW m^{-2}.

The calculated vertical Darcy velocity in the crystalline basment below the Permocarboniferous trough amounts to about $2*10^{-10}$ m/s and at the northern boundary of the trough to about $7*10^{-10}$ m/s, the latter corresponding to a maximum of 20 mm/year.

Discussion

As one of the consequences of the model calculations it must be concluded that, if the pronounced thermal anomaly in northern Switzerland is due to ascending deep groundwater (which seems to be the most plausible mechanism), the vertical permeability of the crystalline basement below or around the Permocarboniferous trough has to be in the range between $5*10^{-15}$ m^2 and $3*10^{-16}$ m^2, depending on the structural position (cf. Figure 5 and Table 1). These values can be characteristic down to depths of 6-7 km.

For this considerable depth the permeability values found seem to be questionably high. Numerous investigations based on drillhole data, isotope studies, petrologic and tectonic evidence have shown, however (see e.g. Costain et al., 1987), that considerable permeability can persist down to this depth in crystalline rocks. Even in tectonically stable regions like the Baltic Shield groundwater flow is present to great depths (Kola drillhole, Kremenetsky and Ovchinnikov, 1986).

The modeling results also raise the question about the origin of water at the depths which are required by the model. A further, possibly related question is why the ascent of water is geographically confined to the area of the lower Aare valley.

The upflow seems to be promoted by the border zones of the Permocarboniferous trough. The origin of the ascending deep groundwater is less clear, especially with respect to the possible zone(s) of infiltration. There are some lines of evidence (geochemical and isotopic signatures) which point towards an origin of the deep crystalline water from the north (Black Forest massif, cf. Kanz 1987) but an origin from the south (i.e. from the Alpine region) cannot be ruled out.

Conclusions

Uprising deep groundwater is the only satisfactory explanation to interpret the pronouncedly positive heat-flow density anomaly, geographycally bound to the Permocarboniferous trough of northern Switzerland.

For a quantitative assessment of flow rate and direction of deep groundwater circulation, numerical two-dimensional, coupled heat and fluid flow modeling revealed pronouncedly upward flow of water in the crystalline basement from depths of about 6-7 km. In the vicinity of the trough, vertical permeabilities in the order of 10^{-15} m^2 prevail in large parts of the basement; Darcy velocities are in the order of 10 mm/y.

Careful model calibration is essential: it is demonstrated that the numerical model successfully fits to a variegated set of field data: hydraulic potential of aquifers, flow rate and temperature of naturally discharging thermal water, subsurface temperature field and surface HFD distribution.

Further, preferably 3-D numerical modeling of coupled heat and mass transfer as well as a thorough analysis of the HFD signature (vertical variation, dependence on heat production of basement rocks) could help to answer the still open questions.

References

Bodmer, Ph., and Rybach, L., Geothermal Map of Switzerland (Heat Flow Density), Beitr. Geol. Schweiz, Ser. Geophysik, Nr. 22, 47 p., 1984.

Bodvarsson, G.S., Mathematical modeling of the behavior of geothermal systems under exploitation. Ph. D. Thesis, Lawrence Verkeley Laboratory, University of California, 353 p., 1982.

Costain, J.K., Bollinger, G.A., and Speer, J.A., Hydroseismicity: A hypothesis for the role of water in the generation of intraplate seismicity. Seism. Res. Let., 58, 41-64, 1987.

Diebold, P., Erdwissenschaftliche Untersuchungen der Nagra in der Nordschweiz; Strömungsverhält-nisse und Beschaffenheit der Tiefengrundwässer. Mitt. aarg. natf. Ges., 31, 11-52, 1986.

Deichmann, N., and Rybach, L., Earthquakes and Temperatures in the Lower Crust below the Northern Alpine Foreland of Switzerland. In: Lower Continental Crust: Properties and Processes, eds. D.M. Fountain, R.F. Mereu, St. Mueller, AGU Monograph., in press, 1988.

Gorhan, H.L., and Griesser, J.-Cl., Geothermische Prospektion im Raume Schinznach Bad - Baden., Beitr. Geol. Schweiz, Geotechn. Ser., Liefg. 76, 73 p., 1988.

Griesser, J.-Cl., Geothermische Prospektion zwischen Baden und Schinznach; Geophysikalische Untersuchungen und thermohydraulische Modellrechnungen, Ph.D. Thesis Nr. 7845, ETH Zurich, 209 p., 1985.

Haenel, R., and Zoth, G., Heat flow measurements in Austria and heat flow maps of Central Europe. Z. Geophys., 39, 425-439, 1973.

Kanz, W., Grundwasserfliesswege und Hydrogeochemie in tiefen Graniten und Gneisen, Geol. Rdschau, 76, 265-283, 1987.

Kremenetsky, A.A., and Ovchinnikov, L.N., The Precambrian continental crust: Its structure, composition and evolution as reveale by deep drilling in the USSR. Precambr. Res., 33, 11-43, 1986.

Laubscher, H., Die tektonische Entwicklung der Nordschweiz, Eclogae geol. Helv., 80, 287-303, 1987.

Matter, A., Faziesanalyse und Ablagerungsmilieus des Permokarbons im Nordschweizer Permokarbon-Trog, Eclogae geol. Helv., 80, 345-367, 1987.

NAGRA, Project Gewähr 1985: Feasibility and safety studies for final disposal of radioactive wastes in Switzerland, Swiss National Cooperative for the Storage of Radioactive Waste, Baden, 75 p., 1985.

NEFF, Forschungsprojekt "Erschliessung geothermischer Vorkommen", Teilprojekt 1: "Nutzbarmachung der Geothermie im Raume Baden und Umgebung", Schlussbericht, Swiss National Energy Research Fund, Basel, 105 p., 1984.

Rybach, L., Eugster, W., and Griesser, J.-Cl., Die geothermischen Verhältnisse in der Nordschweiz, Eclogae geol. Helv., 80, 521-534, 1987.

Schmassmann, H., Neue Erkenntnisse zur Beschaffenheit der Tiefengrundwässer der Nordschweiz, Eclogae geol. Helv., 80, 569-578, 1987.

Simmons, G., Interpretation of heat flow anomalies - 1. Contrasts in heat production, Rev. Geophys. Space Phys., 5, 42-52, 1967a.

Simmons, G., Interpretation of heat flow anomalies - 2. Flux due to initial temperature of intrusives, Rev. Geophys. Space Phys., 5, 109-120, 1967b.

Thury, M., and Diebold, P., Ueberblick über das geologische Untersuchungsprogramm der Nagra in der Nordschweiz, Eclogae geol. Helv., 80, 271-286, 1987.

Contribution no. 580, Institute of Geophysics, ETH Zurich.

HEAT FLOW IN A SEDIMENTARY BASIN IN CZECHOSLOVAKIA: EVALUATION OF DATA WITH SPECIAL ATTENTION TO HYDROGEOLOGY

Vladimír Čermák

Geophysical Institute, Czechosl. Acad. Sci., 141-31 Praha 4, Czechoslovakia

Abstract. Sub-surface temperature fields may be considerably affected by active ground water systems, thereby seriously hampering the interpretation of heat flow data. A simple model of a horizontal aquifer is proposed, which allows a quantitative evaluation of the convective component of heat transfer. This model was applied to geothermal data obtained in two localities in the marginal zone of the Bohemian Cretaceous Basin. It is shown that within a few kilometres from the recharge zone the observed surface heat flow may be underestimated by 50 to 100 per cent and the corresponding hydrogeological corrections were proposed.

Introduction

To determine the heat flow density value, two parameters must be measured, namely, the increase of temperature with depth (temperature gradient) and the coefficient of the thermal conductivity of rocks. If the heat flow value obtained is to be representative of the heat outflow from depth, conductive heat transfer must dominate in the area of measurement. This is usually satisfied but it may be problematic in large sedimentary basins with permeable rocks showing large scale convective heat transport by underground water flows. A difficult situation may arise when both conductive and convective components of heat transfer are of the same order of magnitude. The manifestation of near-surface convection may not be fully realized and a "deep" origin may be ascribed to the observed apparent anomaly, or alternatively, a genuine anomaly may be disregarded due to the distortion of the measured field. Recognition of a convective heat component and the calculation of the corresponding correction are thus of great importance for heat flow studies in many regions.

This paper is a condensed summary of the results published earlier [Čermák and Jetel, 1985]. The original work contributed to the heat flow values measured as a part of the heat flow research programme in Czechoslovakia and focused on the assessment of the possible effect of the underground water movement on the temperature field in a large sedimentary basin located in the north-eastern sector of the Bohemian Massif, the Bohemian Cretaceous Basin.

Tectonic Setting

The Bohemian Massif, part of the Variscan branch of the Hercynian system, represents a large platform-type block in Central Europe (Fig.1). The Bohemian Cretaceous Basin is an elongated sedimentary basin, which separated the most rigid and stable part of the massif from the boundary areas of intensive Variscan tectogenesis to the north. The position of the basin is believed to correspond to an old tectonic suture in the frontal area of the Caledonian range. This zone was subjected to large disturbances as early as during the Precarboniferous sedimentation, when the sediments filled the depressed relief of the crystalline rocks and formed the present basement below a great part of the Cretaceous Basin. In response to the Late Mesozoic Alpine-Carpathian orogenesis, the subsidence and sedimentation of this zone continued. The maximum thickness of the Cretaceous sediments (Cenomanian and Turonian) is 900 metres in the axial part of the basin decreasing laterally in both directions to the basin margins. Hydrogeologically, the Cretaceous Basin, together with the underlying Permocarboniferous sediments, is a large and unique water bearing system. Marginal and inner outcrops of aquifer (mainly sandstones) within the basin are the main recharge areas. The discharge zone of the entire system is located in the valley of the Labe river, which cuts in a zig-zag shape through the basin [Jetel, 1974]. Although the role of the convective heat transport by underground water was suspected in early interpretations of the heat flow pattern in the Bohemian Massif [Čermák, Jetel and Krčmář, 1968], no quantitative arguments were offered to support it.

Copyright 1989 by
International Union of Geodesy and Geophysics
and American Geophysical Union.

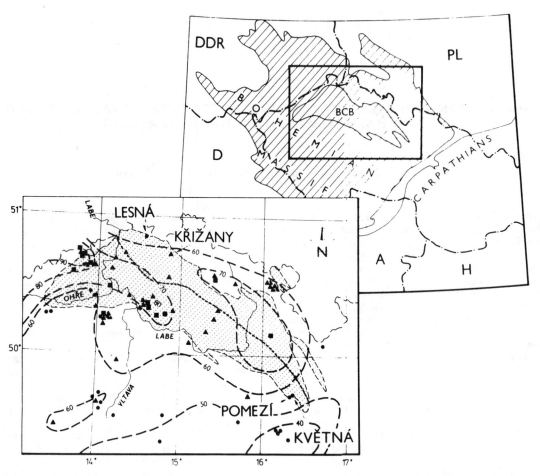

Fig.1. Map showing the location of the studied areas within the Bohemian Cretaceous
Basin. The index map (above) shows the position of the Bohemian Massif and its contact with the Alpine-Carpathian system. Detailed map (below) indicates the Bohemian
Cretaceous Basin (stippled), broken line represents the basin axis. Isolines indicate regional heat flow pattern (mW m^{-2}), heat flow values are shown by dots ($<$60
mW m^{-2}), triangles (60-80 mW m^{-2}) and squares (over 80 mW m^{-2}).

In this paper the degree of the perturbation
of the heat flow field near the recharge areas
of the marginal zones of the basin is evaluated.
Here, a deficit in surface heat flow should be
observed if ground water flow plays a significant
role in the redistribution of the heat flow in
the near-surface rocks. A simple model for an
evaluation of the perturbation of the heat flow
field in a region underlain by a horizontal
aquifer near a recharge area is proposed (Fig.2).
Individual parameters of the model are specified
and the model is applied to observed data. Two
localities are selected for the calibration of
the model. One in the easternmost edge of the
basin consists of two holes in the Pomezí area
(Pomezí and Květná boreholes), the other area,
with Křižany and Lesná holes, is near the basin's
northern limit. The two localities differ in

their local structures, aquifer depths and thicknesses, thus they may characterize the limiting
conditions of the problem.

Regional Distortion of Heat Flow Field
by Underground Water Movement

The model used [Jetel, 1982] comprises a hari-
zontal aquifer bounded above and below by relatively impermeable rocks (Fig.2). It is assumed
that convective heat transfer takes place within
the aquifer through horizontal fluid transport
described by the specific discharge, while across
the boundaries of the aquifer only conductive
heat transfer is possible, and that only vertical
conduction exists in the impermeable rocks above
and below the aquifer.

The temperature distribution is given by the

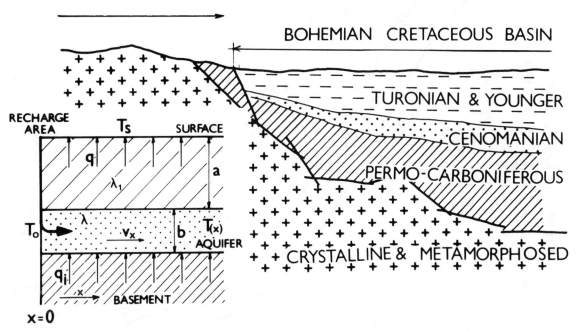

Fig.2. Schematic diagram of the model aquifer system (symbols are explained in the text) together with the typical cross section of the marginal part of the Bohemian Cretaceous Basin.

solution of the partial differential equation [Mytnyk, 1970,1978] :

$$\lambda b \frac{d^2 T}{dx^2} - b \rho c v_x \frac{dT}{dx} + q_i - \frac{\lambda_1}{a} (T-T_s) = 0 \ , (1)$$

where b is the aquifer thickness, $T=T(x)$ is the temperature in the aquifer at a distance x from the recharge zone, ρ is the water density, c specific heat of water, v_x - specific discharge (Darcian velocity), q_i - heat flow from basement rocks (basal or deep heat flow), λ - thermal conductivity of the aquifer rocks, λ_1 - thermal conductivity of the cap rocks, a - depth to the aquifer, T_s - surface temperature.

For the boundary conditions : $T = T_o$ at $x = 0$, and T finite at $x = \infty$, the solution of (1) can be expressed as follows :

$$T = T_r(1 - e^{-nx}) + T_o e^{-nx} \ , \qquad (2)$$

where

$$T_r = T_s + q_i a / \lambda_1 \ . \qquad (3)$$

T_r is the equilibrium aquifer temperature, and T_o is the temperature of the recharging water at $x=0$.

The value of the logarithmic decrement n is given by [Mytnyk, 1970] :

$$n = - \frac{\rho c v_x}{2 \lambda} \left[1 - \sqrt{1 + \frac{4 \lambda \lambda_1}{ba(\rho c v_x)^2}} \right]. \qquad (4)$$

The circulation of water in the aquifer causes a perturbation of the temperature field. The perturbation is attenuated with the increasing distance from the recharge area, and it is smaller for greater values for n.

The surface heat flow (q) above the aquifer is expressed by :

$$q = q_i + (q_o - q_i) \ e^{-nx}, \qquad (5)$$

where

$$q_o = \lambda_1 (T_o - T_s)/a \ . \qquad (6)$$

The observed (surface) heat flow measured above the aquifer differs from the deep heat flow by the amount of heat transferred away by the moving water within the aquifer. The perturbation (q/q_i) was evaluated as a function of the distance from the recharge area for the simplest

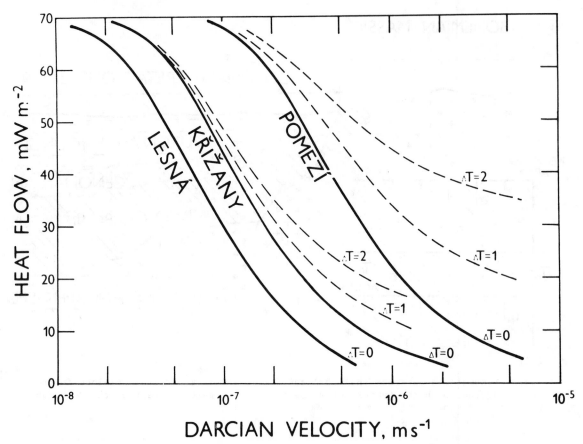

Fig.3. Surface (disturbed) heat flow above the aquifer as a function of the Darcian velocity for constant deep heat flow of 70 mW m^{-2}, Δ T is given in $^\circ$C.

case Δ T=T_o-T_s= 0 (i.e. q_o=0), which corresponds to the rapid infiltration of surface water to the aquifer in the recharge zone [Čermák and Jetel, 1985] . In real cases when slower infiltration results in heating of the descending fluid, ΔT will be greater than zero, and the perturbation of the heat flow according to (5) will be smaller. As the infiltration rate approaches zero, T_o approaches T_r , which results in q = q_i and no hydrogeological perturbation. The case given for Δ T = 0 thus represents the most unfavorable conditions and maximum distortions. The obvious dominant dependence of the perturbation on the Darcian velocity is quite distinct and the hydraulic properties of the aquifer control the heat transport. It was shown that if the Darcian velocity is less than 1x10^{-8} m.s^{-1}, the heat transfer is conductive, the effect of the water flow on the measured heat flow being negligible. For values of fluid velocity between 2 and 5x10^{-8} m.s^{-1} the effect of the ground water movement on the heat flow may be observable in the close neighbourhood of the recharge area up to a distance of 1-2 km, and can be easily corrected. Much greater perturbations, hopefully still cor-

rectable if the parameters are measured with sufficient accuracy, occur for Darcian velocities between 5x10^{-8} and 5x10^{-7} m.s^{-1}. For still higher velocities the perturbation of the temperature field may be so great that a reliable correction is impossible within 10 to 15 km of the recharge zone. The correction is quite uncertain at larger distances. For velocities of the order of 10^{-6} m.s^{-1} , the heat transfer is almost completely convective and the corresponding surface heat flow is dramatically affected at distances as large as 20-50 km from the recharge zone.

Parameter n is not greatly affected by variations in the thermal conductivity of the aquifer (λ) and only slightly affected by variations in the thermal conductivity of the overlying layer (λ_1). Smaller perturbations occur for higher λ_1. If Δ T= 0, q=q_i(1-e^{-nx}) , and the thickness and depth of the aquifer only appear in the value of exponent n. In this specific case, a thinner aquifer at a greater depth causes the same perturbation of the temperature field as a thicker but relatively shallower aquifer.

The Cenomanian aquifer in the Cretaceous Ba-

TABLE I. Coordinates, Heat Flow and Hydrogeological Parameters

Locality	Pomezí	Květná	Křižany	Lesná
Latitude	49°42.8'	49°44.1'	50°55.0'	50°52.0'
Longitude	16°19.7'	16°21.8'	14°52.0'	14°35.0'
Elevation, m	606	572	393	450
Depth Interval, m	50-130	90-170	120-590	350-660
Number of Temperature Data	9	6	48	32
Temperature Gradient, mK m^{-1}	13.9±1.0	14.8±2.7	20.7±0.7	17.7±0.1
Number of Conductivity Data	8	11	7	7
Thermal Conductivity, W m^{-1}K^{-1}	1.84±0.59	1.60±0.31	3.04±1.16	3.28±0.75
Aquifer Thickness, m	28	50	66	37
Depth to Aquifer, m	92	100	535	698
Distance from Recharge Zone, m	2700	5300	4600	1900
Piezometric Gradient	0.033	0.023	0.037	0.067
Hydraulic Conductivity, m s^{-1}	3.9x10^{-5}	4.0x10^{-5}	1.9x10^{-6}	7.5x10^{-6}
Permeability, m^2	5.6x10^{-12}	5.3x10^{-12}	1.9x10^{-13}	7.7x10^{-13}
Darcian Velocity, m s^{-1}	1.3x10^{-6}	9.2x10^{-7}	7.0x10^{-8}	5.0x10^{-8}
Measured Heat Flow, mW m^{-2}	26	24	68	58
Corrected Heat Flow, mW m^{-2}	99	68	89	88

sin ranges in the thickness between 20 and 120 metres and its depth below the Turonian and younger strata varies from 50 to 800 m (Fig.2). Therefore the product b·a ranges from (1 to 70) x10^3 m^2. However, since the margins of the basin are not usually limited by large displacement faults, except in the northern rim, the depth of the Cretaceous sediments in the marginal regions is moderate and the basement slopes gently from the marginal to the axial portions of the basin. Therefore, the maximum perturbation of the temperature field, even for high Darcian velocities, should be limited to the area within a few kilometres of the recharge zone, while a smaller or negligible effect could be expected in the inner parts of the basin. The case of Křižany with b·a =35x10^3 m^2 is one exception, while the case of Pomezí with b·a=2.6x10^3 m^2 appears to be more typical.

The case $\Delta T=0$ is the simplest approximation of the real conditions, thus the general applicability of the above-mentioned conclusions may be limited. In real conditions, T_o does not equal T_s, i.e. q_o is not zero, and the perturbation of the temperature field above the aquifer is controlled, for a constant value of the product b·a, and for a constant unit change of temperature difference ΔT, by the variation of the depth to the aquifer (a) (see eq.6). For a shallower aquifer q_o approaches q_i and hydrogeological perturbations decrease. The model proposed here only considers horizontal and subhorizontal laminar ground water flow. To complete the general picture, one has to consider also the effect of migration of the fluids through rocks of low permeability, i.e. aquitards. This migration may be

upward as well as downward. It is believed that the upward migration may be of greater importance for the heat flow correction because this movement predominates at greater depths. While this effect seems to be negligible in Cretaceous sediments as compared with the lateral components, it may be of some significance in the Permocarboniferous rocks of the basement. Smith and Chapman [1983] predicted a regional distortion of heat flow for permeability values as low as 5x 10^{-16} m^2, which is close to the permeability of Permocarboniferous rocks at the depth of 1 km (5x10^{-17} to 1x10^{-16} m^2), below this depth permeability rapidly decreses [Jetel, 1982].

Proposed Corrections

Figure 3 depicts the uncorrected surface heat flow for the investigated holes as a function of the Darcian velocity. The basal heat flow is assumed to be 70 mW m^{-2} and the temperature of infiltrating water T_o equals the surface temperature T_s, i.e. $\Delta T=0$. As an example, a 20 % reduction in the heat flow in the Pomezí hole, with a depth of 92 m to the aquifer (Tab.I), corresponds to $v_x = 2.3x10^{-7}$ m.s^{-1}. In holes at Křižany and Lesná (with depths to the aquifer of 535 and 698 m), comparable reductions occur at $v_x = 6x10^{-8}$ and 3x10^{-8} m.s^{-1}, respectively. Likewise, a 50 % reduction in the surface geothermal activity corresponds to v_x-values of 5.5x10^{-7}, 1.7x10^{-7} and 9x10^{-8} m.s^{-1} for Pomezí, Křižany and Lesná holes. For positive values of ΔT the effect is less prominent, thus the solid lines in Fig.3 correspond to the maximum values while the broken lines show probable values for cases when the infiltrating water is partly heated before it reaches the aquifer. The deeper the aquifer the more likely such heating is to occur.

Taking into account the probable Darcian velocities for all locations based on hydrogeological studies, we evaluated the most reasonable heat flow corrections. Even though the corrections proposed (Table I) are rather possible limits than exact numbers, they give a better insight into the regional distortion of the heat flow pattern. The corrected values are those most probable of a whole spectrum of possible ones [Čermák and Jetel, 1985]. The corrected heat flow for the Pomezí hole (99 mW m^{-2}) is somewhat high and value of about 70 mW m^{-2} would seem to be more realistic with regard to the regional conditions. This fact may be explained by the uncertainty of the Darcian velocity estimate. Assuming values of the heat flow of about 70 mW m^{-2}, the range of the required v_x values is 7 to 9x 10^{-7} m.s^{-1} (Fig.3). In the Květná hole, the estimates of the q_i values, based on stratigraphy [Čermák and Jetel, 1985] were evidently too low; more reasonable results were obtained for the model where the position of the aquifer was adapted to fit the measured temperature-versus-depth curve and the aquifer was identified with the layer of the minimum temperature gradient between 50 and 100 m depth and $v_x = 9.2x10^{-7}$ m. s^{-1}. In the Křižany hole the assumption of $v_x = 7x10^{-8}$ m.s^{-1} leads to the corrected heat flow of 89 mW m^{-2}, which may be also a little too high. However, values of q_i between 80 and 90 mW m^{-2} are not inconsistent with the assumption of a deep crustal structure beneath the northeastern continuation of the Ohře rift in this area [Čermák and Šafanda, 1982]. This would also apply to theLesná hole, where the Darcian velocity values 2 to 5x10^{-8} m.s^{-1} yielded corrected values of heat flow of 63 to 88 mW m^{-2}.

Conclusions

(1) By applying a simple model of lateral convection it was shown that in the marginal portions of the Bohemian Cretaceous Basin the ground water movement might cause a decrease of the observed surface heat flow by up to several tens of per cent.
(2) Near the northern tectonic margin of the basin, lateral movement of relatively cold recharge water descending to a depth of 600-700 m may affect the temperature field up to a distance of 5-15 km from the marginal faults. At greater distances from the recharging area, heat flow is less distorted.

(3) No considerable distortion of surface heat flow occurs in the Cretaceous Basin if the ground water Darcian velocities are less than 1x10^{-8} m. s^{-1}. This appears to be generally valid for other sedimentary basins of a similar depth, where water flow is approximately horizontal.

Acknowledgements. The author wishes to express his sincere thanks to Dr Ján Jetel of the Geological Survey of Czechoslovakia who collaborated in the preparation of the original paper and who made most of the hydrogeological interpretation. A number of colleagues Dr S Bachu, Prof A E Beck, Prof D S Chapman, Prof Y Eckstein, Dr F Horváth, Dr A M Jessop, Prof L Rybach read this manuscript and proposed critical comments and additional remarks and thus helped in its improving and final preparation.

References

Čermák, V., Jetel, J., and Krčmář, B., Terrestrial heat flow in the Bohemian Massif and its relation to the deep structure, Sb.geol.ved, UG, 7, 25-41, 1968.

Čermák, V., and Jetel, J., Heat flow and groundwater movement in the Bohemian Cretaceous Basin (Czechoslovakia), J.Geodynamics, 4, 285-303, 1985.

Čermák, V., and Šafanda, J., Mapa tepelného toku na území Československa, Open file report, Ústřední ústav geologicky, Praha, 12 pp.,1982.

Jetel, J., Hydrogeologie podloží české křídové pánve, In: M.Malkovský (Ed.), Geologie české křídové pánve a jejího podloží, pp.156-190, Ústřední ústav geologický, Praha, 1974.

Jetel, J., Hydrogeologické vlastnosti hornin v geotermálně perspektivních oblastech Českého masívu v hloubkách 1-6 km. In: T.Pačes (Ed.), Hydrogeologické aspekty využití zemského tepla suchých hornin v perspektivních oblastech ČSR, pp. 8-21, Ústřední ústav geologický, Praha, 1982.

Mytnyk, M.M., Stacionarnyi teplovoy rezhim gorizontalnogo vodonosnogo plasta. In: Problemi gidrogeologiyi i inzhenernogo gruntovedeniya, pp.253-269 (in Russian), Naukova Dumka, Kiev, 1970.

Mytnyk, M.M., Analiticheskiye metodi rascheta stacionarnogo teploperenosa i ikh ispolzovaniye v gidrogeologiyi. In: V.M.Lyalko and M.M.Mytnyk (Eds.), Issledovaniya processov perenosa tepla i veschestva v zemnoy kore, pp.23-49 (in Russian), Naukova Dumka, Kiev, 1978.

Smith, L., and Chapman, D.S., On the thermal effects of groundwater flow, 1. Regional scale systems, J.Geophys.Res., 88, 593-608, 1983.

THERMAL EFFECT OF HYDROGEOLOGY IN CLOSED BASINS

Lajos Stegena

Eötvös University, H-1083 Budapest, Kun B. Tér 2.

Introduction

The subsurface water-budget of younger sedimentary basins differs essentially for open and for closed ones. The open basins are open hydrologically toward the sea; examples of this kind could be the Aquitanian as well as the Po basin, having discharge into the Atlantic and the Adriatic sea respectively. Other basins have no hydrological connection outwards; in the closed basins the infiltrated meteoric water and the connate water expelled by compaction of the sediments have to return to the surface of the basin. A typical example of these closed basins is the Pannonian basin.

In order to estimate the geothermal effect of the migrating subterranean water, a quantitative description of the water's circulation pattern in the basin is needed. A basic quantitative description of the Aquitanian basin has been given by Besbes et al. (1978). The present study deals with the hydraulics and their geothermal effects of a closed basin taking the Pannonian basin as a typical example.

The problem is handled in two steps. First, in a regional scale model a hydraulic homogeneity is supposed for the whole basin; the bulk permeability depends only on the depth. This model is far from the reality because the subsurface flow takes place mostly along zones of weakness, i.e. linearly and not areally. It gives however a good picture concerning the contribution of water migration to the heat flow in a regional scale, for the whole basin. In a second step the thermal effect of water moving along breaks and zones of weakness are discussed.

Water Migration and its Geothermal Effect: a General Model in Regional Scale

The mean value of HFD on the Pannonian basin amounts to 106 ± 15 mWm^{-2} (Dövényi et al., 1983). The aim of this chapter is to determine: how

Copyright 1989 by
International Union of Geodesy and Geophysics
and American Geophysical Union.

this high value is influenced by subsurface water movements. In order to answer this question we need to know the rate of flow of the in- and outfiltrating water and the depth of its penetration.

Suppose a steady-state for the water migration, the recharge and discharge rates for the whole basin are the same. This supposition is not going without saying: at large distances, low hydraulic gradients and low permeabilities, the relaxation times are very large, in geological sense too (Tóth, 1980).

The amount of moving groundwater rate for the whole Pannonian basin was determined by Böcker and Liebe (1980) as 35 m^{3}s^{-1}. This is the recharge rate and also the discharge rate for the basin. This value is based on the 1:0,5 M map of equilibrium water levels in some thousand boreholes; after correcting these levels by thermal expansion of water the hydraulic gradients were calculated. The permeabilities were based on pumping experiments in boreholes and on laboratory measurements. In spite of the fact that on several places geological analogies only were at disposal, the derived value seems to be reasonable.

The whole basin of about 10^5 km^2 proportions in 59 % recharge and 41 % discharge area. This, and the distribution of recharge (R) and discharge (D) area is based on equilibrium water levels (Erdélyi, 1985; Fig. 1.).

In order to estimate the thermal effect of the filtrating water, the vertical distribution of the water flow is required. The depth of penetration can be estimated with the aid of permeabilities. Fig. 2. shows the result of a hundred permeability determinations for the Pannonian sediments (Á. Szalay, personal communication, 1987). The basin's young-Tertiary sediments consist of 50 % shale and 50 % sandstone on the average and this proportion does not change essentially with depth. Consequently the solid curve of Fig. 2 was used for the further calculations, which curve was derived for every depths as follows

$$k = \frac{0,5k_{SA}+0,5k_{SH}}{k_{SA} \cdot k_{SH}}$$

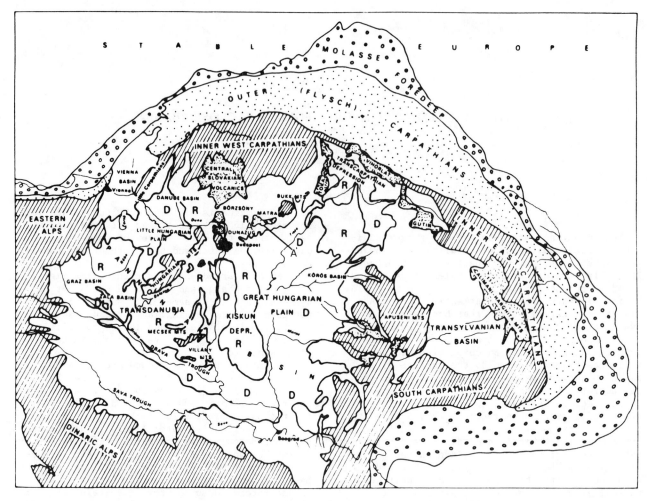

Fig. 1. Geographic and geologic main units of the Carpatho-Pannonian region and the recharge (R) and discharge (D) areas (Stegena, 1985).

(k - permeability, SA - sandstone, SH - sandstone, SH - shale). It is supposed that this averaged curve gives a satisfactory approximation, at least for our purposes.

Suppose the obliqueness of possible infiltration paths is independent of depth, the distribution of the total 35 m^3s^{-1} rate for various penetration depths is proportional to the vertical hydraulic resistance, h/k (h: depth of penetration, k: mean permeability from o to h) (Fig. 2).

Fig. 3 upper part shows the result of the calculation: the penetration depth for 32550 ls^{-1} from the total 35000 ls^{-1} (99,5 %) is less than 0,5 km and the rate of water penetrating between 4-5 km gives only 3 ls^{-1} (0,009 %).

Based on geothermal temperature maps (e.g. Dövényi et al., 1983) the heat flow density contribution of the down and upward mowing fil-

tration of the down and upward mowing filtration water can be calculated, regionally for whole basin (Fig. 3, upper part):

$$HFD = \pm \frac{r_h \cdot c \cdot T_h}{Re \text{ or } Di}$$

Where r_h the rate of water flow for every depth interval, c the specific heat of water, T_h the mean temperature difference between the depth interval and the surface. On recharge areas (Re= $0,59.10^5$ km^2) the heat flow density contribution HFD is negative, on discharge areas (Di=$0,41$. 10^5 km^2) HFD is a positive value.

The effect of the uppermost 500 m was not taken into consideration, because geothermal measurements are always made at deeper depths. Because of this, the penetration into the uppermost 500 m (which is very lage, 32550 ls^{-1}) is

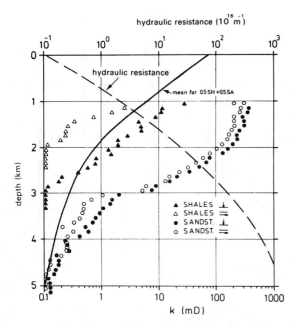

Fig. 2. Permeability values of the Pannonian basin's sediment as a function of depth (Á. Szalay, personal communication, 1987), the mean curve (solid line) used for the present calculations, and the mean hydraulic resistance (dashed line). Permeabilities parallel and perpendicular to the layering are given separately.

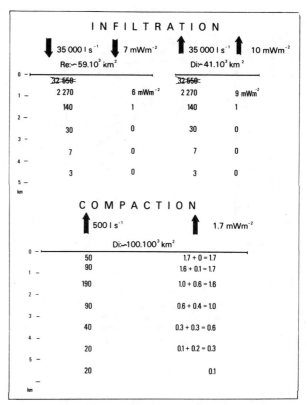

Fig. 3. Regional water-budget of the infiltration and compaction waters of the Pannonian basin and their geothermal effect. For the calculation homogeneously distributed, areal water migation is supposed.

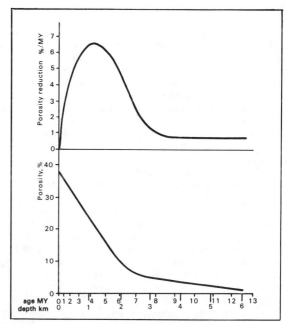

Fig. 4. Average porosity vs. depth and age function for the sediments of the Pannonian basin and the calculated change of porosity with time.

crossed out on Fig. 3. The layers however beneath the uppermost 500 m are thermally not very affected by regional ("areal") water filtration: the mean effect in heat flow density is −7 and +10 mWm^{-2} respectively.

The thermal effect of water expelled by compaction and filtrating upward (Fig. 3, lower part) has been calculated using the mean porosity vs. depth function of the Pannonian basins's sediment (Fig. 4). Knowing the average age of sediments, the temporal change of porosity (%/MY) was calculated. Supposing that the porosity reduction is the consequence of compaction only − for the young Tertiary sediments of the Pannonian basin an acceptable hypothesis − the present-day mean rate of the water expulsion with respect to depth is calculated (Fig. 3, lower part). Using again the geothermal temperature maps, the average contribution of the migrating compaction water to the heat flow density has been calculated. The calculation is similar to the infiltration water but the expulsion water moves upward in the whole basin and the HFD contribution is always positive:

$$HFD = + \frac{r_h \cdot c \cdot T_h}{Re + Di} = \frac{r_h \cdot c \cdot T_h}{10^5 \; km^2}$$

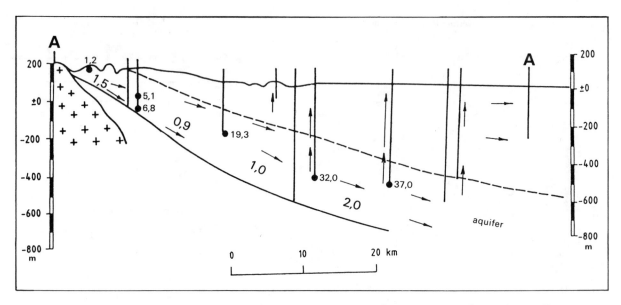

Fig. 5. A section at the southern piedmont of the Mátra Mountain (Fig. 1, A-A').
Small numbers: age of deep waters determined by radiocarbon, in 10^3 years.
Bigger numbers: velocity of water filtration, in m.year^{-1}, calculated from the
radiocarbon ages. Data after Deák, 1980.

The total effect is only 1,7 mWm^{-2} in average, which lies below the accuracy of heat flow determinations.

Geothermal Effect of Subsurface Water Migration Along Belts of Weakness

The waters moving in the depths are not distributed equally, homogene ously, areally in the whole basin as handled previously but are concentrated along belts of hydraulic weakness, fractures etc. The convected heat calculated previously is also concentrated on these zones.

For the local water circulation and its thermal effect in the Pannonian basin, some examples are published (e.g. Horváth et al., 1981, Alföldi et al., 1985, Stegena, 1985). Here two other examples are presented.

Fig. 5 shows a section (A-A' on Fig. 1), at the southern piedmont of the Mátra mountain. The groundwater flows basinwards, its age determined by 14C measurements in boreholes. From these data, the Darcy velocities were calculated (1,5 0,9 1,0 2,0 m.year^{-1} as shown on Fig. 5). Taking into consideration the numerical parameters of the aquifer (porosity=20 %, thickness=350 m, gg= 0,05 km^{-1}) and the mean value of the velocity (0,14 m.year^{-1}), the downgoing water gives rise to a heat flow density decrease of 12 mWm^{-2} on average, in the piedmont area. Namely, the water

supply flowing through along the section in a 1 m belt perpendicular to it: $1.350.1,4.0,2$ m^3. year^{-1}=3,2 gs^{-1}. As gathered from Fig. 5, this supply deepen s 18.10^{-3} m for every 1 m path, which corresponds to $18.10^{-3}.0,05=0,9.10^{-3}$ K warming up. Thus, the heat drawn of for every m^2:

$$HFD=3,2.0,9.10^{-3}.1 \text{ cal m}^{-2}s^{-1}=12 \text{ mWm}^{-2}.$$

Another example is presented on Fig. 6, which features the Transdanubian Mid Mountain, the area between the Danube and Lake Balaton, extended to 7200 km^2. The mean heat flow density measured is 60 mWm^{-2}. Around the mountain, 33 springs originate at the piedmont (Gölz, 1982), with a total thermal output of 320 MW over the air temperature, representing probably all the thermal output by water flow. This is equivalent to a heat flow density of 44 mWm^{-2} over an area of 7200 km^2. Adding the two values, we get 104 mWm^{-2}; this coincides with the mean (106 mWm^{-2}) of the Pannonian basin. Then, the heat flow density beneath the sediments has to be the same as in the basin.

There are places however in the Pannonian basin, where the geothermal conditions are strongly disturbed by unknown hydraulic processes. In some cases, it is not possible to take into consideration these disturbances with the necessary accuracy.

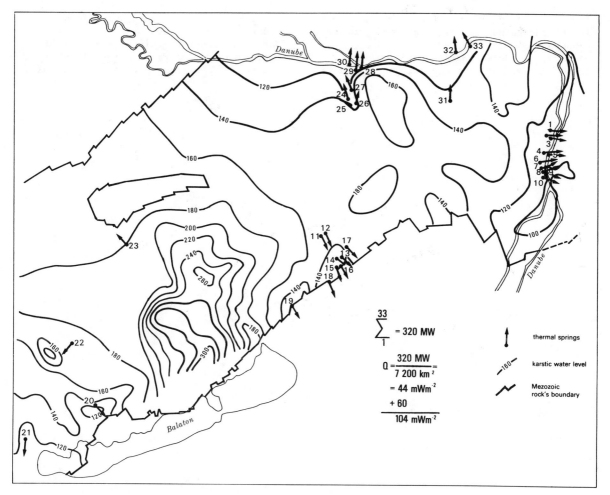

Fig. 6. The Transdanubian Mid Mountain with the 33 thermal springs at its piedmont (Gölz, 1982). The mountain is situated between the Danube and Lake Balaton, see Fig. 1 for orientation. The total thermal output of the springs gives 320 MW, which is equivalent to a mean HFD of 44 mWm^{-2} for the area. The measured mean of the mountain is 60 mWm^{-2} and 106 mWm^{-2} is the mean for the Pannonian basin.

References

Alföldi, L., Gálfi, J. and Liebe, P., Heat flow anomalies caused by water circulation. In: L. Rybach (ed.), Heat flow and Geothermal Processes. Journal of Geodynamics, 4, 199–217, 1985.

Besbes, M., Marsily, G. and M. Plaud, Bilan des eaux souterraines dans le Bassin Aquitain. Ann. Inst. Geol. Publ. Hung. LIX., (Hydrogeology of Great Sedimentary Basins.), 1–4, 293–303, 1978.

Böcker, T. and Liebe, P., Exploitation possibilities of karstic stratum water. (In Hungarian). Internal Report of VITUKI, Budapest, 1980.

Deák, J., Environmental isotopes and water chemical studies for groundwater research in Hungary. Isotope Hydrology, 221–249, 1980.

Dövényi, P., Horváth F., Liebe, P., Gálfi J. and Erki I., Geothermal conditions in Hungary. Geophysical Transactions, Budapest, 29, 1, 3–47, 1983.

Erdélyi, M., Geothermics and the deep flow-system of the Hungarian basin. In: L. Rybach (ed.), Heat flow and Geothermal Processes. Journal of Geodynamics, 4, 321–330, 1985.

Gölz, B., Natürliche Warmeleistung der Quellen in Transdanubian Mittelgebirge. (In Hungarian, with German abstract). Földrajzi Értesítő, Budapest, XXXI., 4, 427–447, 1982.

Horváth, F., Dövényi, P. and Liebe, P., Geothermics of the Pannonian basin. Earth Evolution Sciences, 3-4, 285-291, 1981.

Stegena, L., On the subterranean hydraulics of closed basins. In: Geothermics, Thermal-Mineral Waters and Hydrogeology. Theophrastus, Athens, 59-69, 1985.

Tóth, J., Cross-formational gravity-flow of grundwater. In: Problems of Petroleum Migration, AAPG Studies in Geology, 10, 121-167, 1980.

GROUNDWATER FLOW AND GEOTEMPERATURE PATTERN

Ji-yang Wang and Liang-ping Xiong

Institute of Geology, Academia Sinica, Beijing, China

Abstract. Thermal and hydrological data have been obtained and analyzed for the Liaohe, North China and Longyan Basins, and for the Zhangzhou and Beijing geothermal fields. The results, presented as isotherm and iso–water–head maps, exhibit generally a variation of temperature and heat flow along the direction of water flow. For a small basin, such as the Liaohe or Longyan Basin, where the water flow is strong, the variation is exponential and may be explained by a simple analytic model assuming horizontal flow of water in a confined aquifer. For a large scale basin or a geothermal field, such as the Beijing Geothermal Field, we use both a finite element and a finite difference convective model. Good agreements have been obtained. A case study has also been conducted in the Xinzheng Basin to isolate the "conductive" component of heat flow in a recharge zone where the water movement is presumably vertical. A representative value for the region is deduced.

Introduction

Movement of groundwater in the upper–most part of the crust has been known to affect the distribution of temperature and heat flow. In an artesian basin, geotemperature generally increases gradually along the direction of water flow with negative geothermal anomaly appearing in the recharge area due to the infiltration of cold water, and with positive anomaly appearing in the discharge zone due to the upward migration of warm water. On the other hand, if hot water of deep circulation origin invades an aquifer through faults or fracture zones, temperature will decrease along the flow direction with positive geothermal anomaly centers around the invasion zone. In either case, however, the extent and the buried depth of an aquifer, and the velocity of water flow may be traced by using temperature measurements obtained in boreholes. Conversely, it is possible to model the distribution of temperature and heat flow, making use of available hydrological data (e.g., Parsons, 1970; Domenico and Palciauskas, 1973; Boyle, 1979; Mansure and Reiter, 1979; Torrance and Chen, 1980; Stegena, 1982; Smith and Chapman, 1983).

Modeling of the coupled thermal and hydrological regimes is a complex problem. The direction and magnitude of water flow, the aquifer type (confined, unconfined, or semi–confined), the degree of continuity between aquifers, and the thermo–physical properties of the aquifer and aquitard are all factors that affect the pattern of temperature and heat flow observed at the earth's surface. In this work, we focus mainly on the relatively simple problem of thermal effects due to water flow in confined horizontal aquifers.

Mathematical Model

In a groundwater basin, temperature and its gradient may vary along the direction of water flow owing to the exchange of heat between the aquifer and the surrounding rocks. The equation of heat transfer for a confined aquifer is

$$\frac{\partial}{\partial x}(\lambda m \frac{\partial T}{\partial x}) + \frac{\partial}{\partial y}(\lambda m \frac{\partial T}{\partial y}) - \frac{\partial}{\partial x}(m \rho_o C_o v_x T)$$
$$- \frac{\partial}{\partial y}(m \rho_o C_o v_y T) + \Delta q_z = m \rho C \frac{\partial T}{\partial t} \qquad (1)$$

where $\rho_o C_o$ is the specific heat capacity of water, $m = z_2 - z_1$ is the thickness, and λ and ρC the thermal conductivity and specific heat capacity of the aquifer, and $\Delta q_z = q(z_2) - q(z_1)$ the net vertical heat flow across the aquifer. v_x and v_y are the lateral components of water velocity given by

$$v_x = - K \frac{\partial H}{\partial x}$$
and
$$v_y = - K \frac{\partial H}{\partial y} \qquad (2)$$

where K is the hydraulic conductivity and H the water head. The equation for the overlying aquitard is

$$\frac{\partial}{\partial x}(\lambda_r \frac{\partial T}{\partial x}) + \frac{\partial}{\partial y}(\lambda_r \frac{\partial T}{\partial y}) + \frac{\partial}{\partial z}(\lambda_r \frac{\partial T}{\partial z})$$
$$= \rho_r C_r \frac{\partial T}{\partial t} \qquad (3)$$

where λ_r and $\rho_r C_r$ are the thermal conductivity and specific heat capacity of the aquitard.

In general, the system of equations (1) and (3) must be solved using numerical methods (e.g., Tsang et al., 1981; Zhang and Xiong, 1986). However, for a horizontal aquifer with both the thermal and hydrological regimes in steady state, a simple analytical approximation for the aquifer temperature is possible:

$$T = (T_a + q^* \frac{\Delta z}{\lambda})(1 - e^{-nx}) + T_o e^{-nx} \qquad (4)$$

Copyright 1989 by
International Union of Geodesy and Geophysics and American Geophysical Union.

Here, Δz is the thickness of the aquitard, T_a the mean surface temperature, q^* the heat flux at the base of the aquifer, T_0 the water temperature at the inlet of the aquifer, and n a constant given by

$$n = \frac{\rho_0 C_0 v}{2\lambda}\left[1 - \left[1 + \frac{4\lambda\lambda_r}{(\rho_0 C_0 v)^2 m \Delta z}\right]^{\frac{1}{2}}\right] \quad (5)$$

When v is small, i.e., when heat conduction dominates, n may be approximated by

$$n = -\left[\frac{\lambda_r}{\lambda m \Delta z}\right]^{\frac{1}{2}}$$

On the other hand, if v is large, convection dominates and equations (4) and (5) simplify to

$$T = T_r + (T_0 - T_r)\, e^{-nx} \quad (6)$$

and

$$n = -\frac{\lambda_r}{\rho_0 C_0 v m \Delta z} \quad (7)$$

where T_r is the equilibrium temperature at $x = \infty$. Equation (6) shows that for a horizontal aquifer with strong water flow, temperature varies exponentially along the flow direction. This is in agreement with data obtained in small basins of this study.

Case Studies

We present as isotherm and iso– water–head maps, the results of our analyses for thermal and hydrological data collected in different areas of North and Southeast China. For small basins, such as the Liaohe and Longyan Basins, the simple exponential model of equations (6) and (7) applies. For large basins, such as the North China Basin, and for geothermal fields, such as the Beijing Geothermal Field, a full numerical solution for equations (1) and (3) has to be used.

Liaohe Basin

Liaohe Basin in North China is an artesian basin with two aquifers in the Neogene formation. Because of the relatively small extent, "cold" groundwater moves at a high rate from the recharge area in the periphery highlands toward the discharge area in the Liaodong Gulf, resulting in a regional negative geothermal anomaly (Wang and Wang, 1986; Wang et al., 1986). Table 1 summarizes the geothermal gradients obtained in the Neogene and Quaternary sediments. The measured temperature at a depth of 1000 m is about 35°C, and the mean geothermal gradient is 25 mK m^{-1} (Fig. 1). There is an exponential increase of geothermal gradient along the direction of water flow as shown in Fig. 2, in agreement with the simple model of equations (6) and (7).

North China Basin

North China Basin is a large–extent groundwater basin with negative thermal anomaly (Fig. 3) observed only in the foothills of Taihang Mt. to the west and Yanshan Mt. to the North, which are the recharge zones. Away from the recharge areas, temperature and its gradient increase very gradually, and in the central part of the basin, no geothermal anomaly is observed. As discussed by Xiong and Gao (1982), Xiong et al. (1985),

TABLE 1. Summary of mean geothermal gradients (in mK m^{-1}) in Neogene and Eogene formations, Liaohe Basin, North China

	Mean Geothermal Gradient				
Borehole	Q+N$_m$	N$_g$	E$_d$	Es$_{1,2}$	Es$_{3,4}$
G 1–3–9	21.7			30.2	47.6
G 1505	22.6	22.6	33.0	35.4	41.4
CG 1	22.7	24.5	21.1	30.7	53.0
SG 102	23.8	24.1	35.1	40.0	56.4
S 71	29.6	24.2	29.2	32.3	55.1
S 14	29.8	17.9		36.7	53.5
S 2–6–07	20.0	20.0	26.7	35.3	46.0
S 1–3–5	35.4	28.8		40.7	51.9
L 11	24.5	27.7	35.3	39.4	31.4
XJ 1	21.2	21.2	39.7	52.9	
XB 1	18.1	18.1	38.0	43.6	45.0
D 67	32.8	33.0		41.0	43.0
D 87	38.3	39.4		43.3	45.3
D 55	37.3	31.5	38.0	43.7	34.8
D 52	35.6	28.1		47.1	49.8
D 65	31.7	26.6	30.9	31.1	38.5
D 70	40.9	28.8			57.3
DG 4	18.7	18.7	26.7	38.6	44.5
DG 38	22.7	22.7	31.3	45.4	49.4
DG 37	22.9	22.9		45.4	49.9
SA 23	24.2	24.2	36.0	48.1	41.4
SA 91	21.7	21.7	44.7	56.9	
SS 3	30.4	30.4	29.4	36.2	31.2
Q 2–20–7	20.7	20.7	36.7	40.8	
H 4	28.7	20.6	45.1	45.6	55.1
J 89	31.7	31.7	34.8	43.8	
J 93	26.0	26.0	25.4	45.6	40.1
J 91	38.0	26.4	32.2	41.7	46.4
J 92	34.6	34.6	36.6	50.0	44.6
J 88	33.2	32.0	23.3	41.7	45.1
J 94	35.3	31.5	45.0	47.6	38.9
J 87	35.8	25.6	32.7	40.4	41.4
J 203	28.4	28.4	32.4	41.4	42.9
J 117	21.9	21.9	45.9	54.3	
J 83	30.7	30.7	33.3	37.2	49.8
W 7	29.3	29.3	33.1	40.4	42.8
HU 10–11	25.4	25.4	30.9	30.8	
K 1	28.6	23.4	33.3	35.9	40.0
T 6	25.5	23.1	41.4	41.7	44.0
JI 3	31.0	31.0	32.1	32.3	36.7
LO 24	21.6	28.5	32.6	41.0	43.1
N 11	21.6	21.6	30.8	33.5	

Q+N$_m$ Quaternary sediments and Neogene Minghuazheng formation
N$_g$ Neogene Guantao formation
E$_d$ Eogene Dongying formation
Es$_{1,2}$ 1st + 2nd sections of Eogene Shahejie formation
Es$_{3,4}$ 3rd + 4th sections of Eogene Shahejie formation

and Xiong anf Zhang (1988), the structure form and the relief of basement rock are the principal factors that determine the temperature and heat flow patterns in the North China Basin. Groundwater flow appears to play only a secondary role except in the recharge areas.

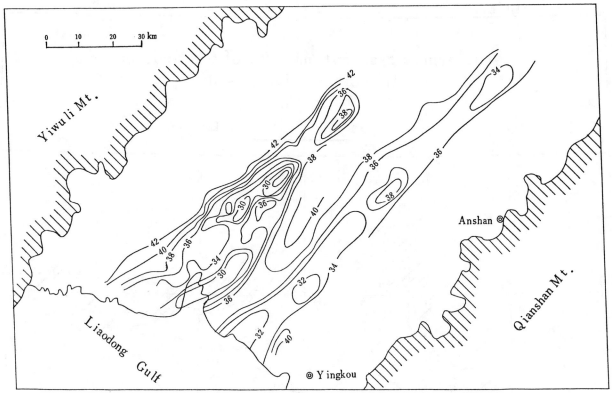

Fig. 1. Isotherm map (in °C) at 1000 m depth in Liaohe Basin, North China.

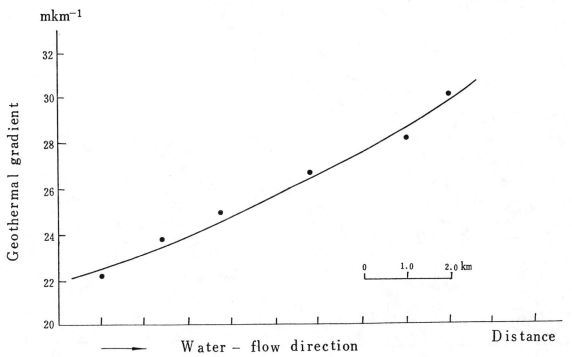

Fig. 2. Variation of mean geothermal gradient (in mK m⁻¹) along water flow direction in Liaohe Basin.

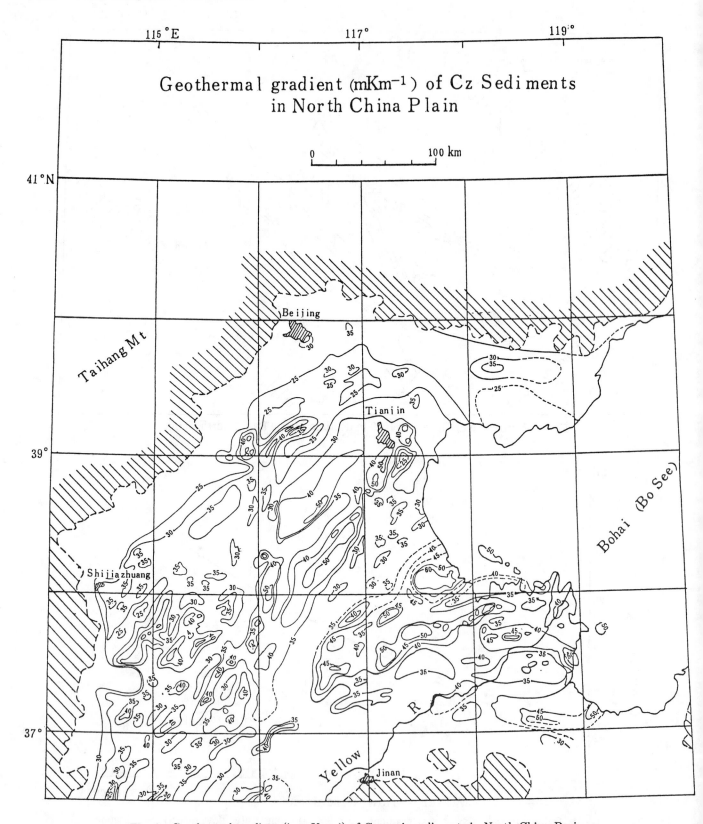

Fig. 3. Geothermal gradient (in mK m⁻¹) of Cenozoic sediments in North China Basin.

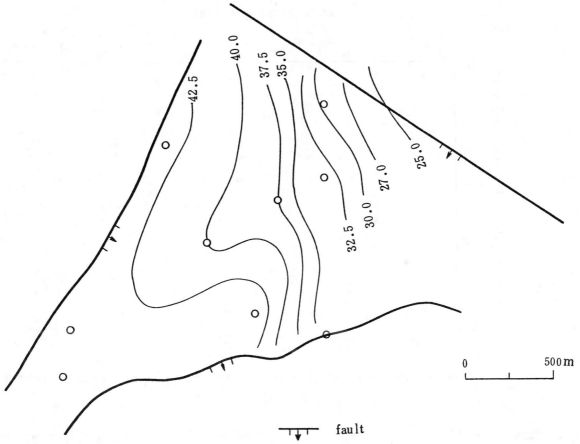

Fig. 4. Isotherm map (in ºC) at 300 m depth in rock strata overlying the Permian aquifer in Longyan Basin of Fujian Prov., SE China. o = temperature measurement borehole.

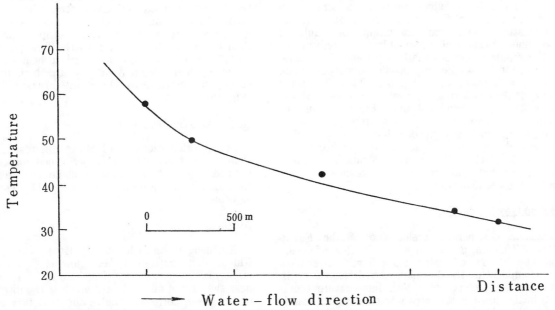

Fig. 5. Temperature variation (in ºC) of "hot" water within the Permian limestone aquifer in Longyan Basin.

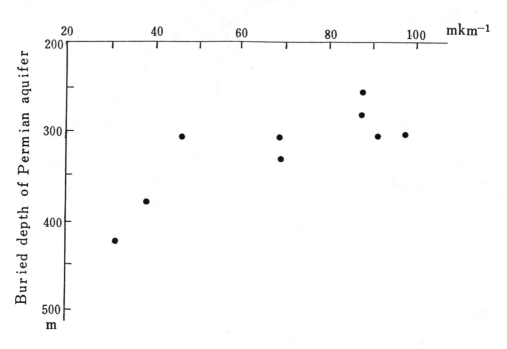

Fig. 6.   Variation of geothermal gradient (in mK m⁻¹) with buried depth of Permian aquifer in Longyan Basin.

Longyan Basin

Longyan Basin of Fujian Province, SE China is a basin of faulting–depression origin. The isotherm map of Fig. 4 shows clearly an enhancement of temperature in the limestone aquifer of Permian age due to the invasion of "hot" water of deep circulation origin along faults or fracture zones. Because of the small lateral extent of this basin, the simple model of equations (6) and (7) applies: Temperature in the aquifer decreases exponentially along the direction of water flow (Fig. 5).   There is also a tendency for the geothermal gradient to decrease with the buried depth of the Permian aquifer (Fig. 6). Thus the Permian aquifer may be treated as a simple heat source in interpreting the regional distribution of heat flow (Xiong and Wang, 1988). The variation of geothermal gradient with depth in the overlying strata is shown in Fig. 7.   The trend again reflects the presence of "heat source" in the Permian aquifer.

Zhangzhou Geothermal Field

Zhangzhou Geothermal Field, also of the Fujian Province, SE China, is a basin where the flow of "cold" groundwater is confined mainly along fissures at shallow depth. Complicating the thermal and hydrological fields, however, is the presence of a high temperature center (Fig. 8) in the down stream area where the water head is low (Fig. 9). This is due to the rise of deep circulation

"hot" water along faults and/or fracture zones. The dynamics of a fault zone controlled geothermal reservoir has been discussed by, e.g., Kassoy and Zebib (1978) and Goyal and Kassoy (1980). Here, the mixing of "hot" and "cold" waters creates an asymmetry in the temperature distribution around the high temperature center. There is a much more rapid temperature drop to the northwest than to the southeast owing to the flow pattern of the "cold" water. The measured surface heat flow value is extremely high (359 mW m⁻²) at the high temperature center but decreases to 69 mW m⁻² in the NW direction (Fig. 10).

Beijing Geothermal Field

Model calculations using finite difference and finite element methods have been performed for the Beijing Geothermal Field. Both the borehole measurements and the model results are presented in Fig. 11. There is good agreement.

Xinzheng Basin

Xinzheng Basin is located in Henan Province, North China.   The pattern of heat flow is shown in Fig. 12 (Deng, 1988). A borehole (Borehole 911) has been drilled near the axis of an anticline, which is the discharge area of the basin, corresponding to the heat flow high. The geotherm obtained in this borehole is shown in Fig. 13.

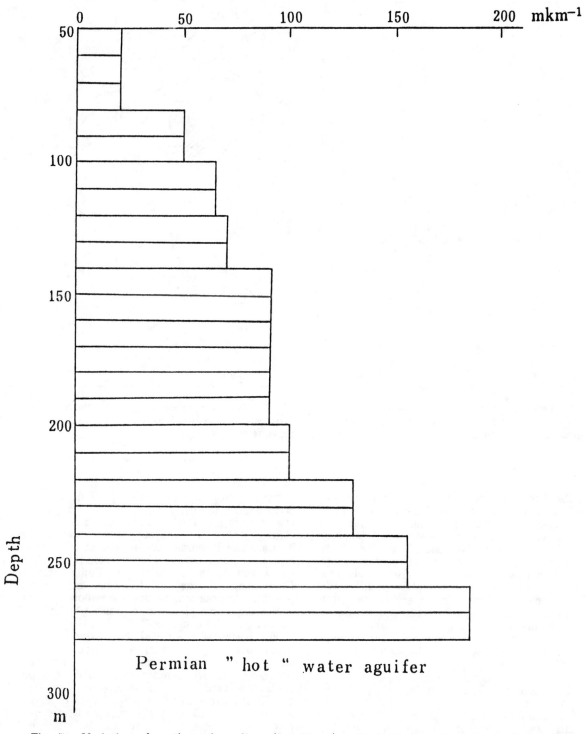

Fig. 7. Variation of geothermal gradient (in mK m⁻¹) with depth in rock strata overlying the Permian aquifer in Longyan Basin.

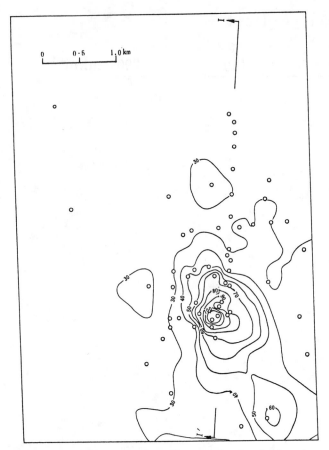

Fig. 8.   Isotherm map (in °C) at a depth of 50 m in Zhangzhou Geothermal Field of Fujian Prov., SE China. o = temperature measurement borehole.

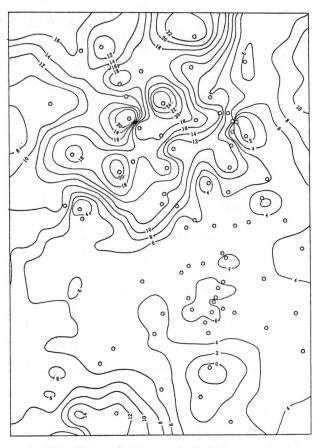

Fig. 9.   Iso—water—head map (in m above sea level) in bedrock of Zhangzhou Geothermal Field. o = water head measurement borehole.

Between the depths of 300 m and 400 m, it is convex, indicating an upward migration of thermal water in the aquitard between the Mid Ordovician (O_2) limestone aquifer and the Upper Carboniferous (C_3) aquifer. Assuming that the water movement is vertical, it is possible to estimate both the water velocity and the conductive component of heat flow (Bredehoeft and Papadopulos, 1965) from the temperature profile. We find a flow velocity of 7.5 m s^{-9}, giving a convective heat flow of 38 mW m^{-2}, which is 37% of the total measured heat flow (102 mW m^{-2}) at the Earth's surface. The conductive component of 64 mW m^{-2} is representative of the region studied, thus supporting the assumption and method used.

Discussion and Conclusion

One of the main objectives in geothermics is to obtain representative background heat flow densities for the regions studied. However, the thermal regimes of many sedimentary basins and geothermal fields are usually perturbed by groundwater movements, making it necessary to carefully identify and remove such perturbations (Sorey, 1971; Kilty and Chapman, 1980). Case studies presented in this work show that in North China, pore—water and/or fissure—water aquifers of Tertiary and Paleozoic age are present in several basins of large or samll extent. These include the Liaohe, North China and Xinzheng Basins. In Fujian Province, SE China, small basins with Permian aquifers are also present. In addition, fissure water flow is widespread over a large area of Mesozoic granitic bodies, including the Longyan Basin and the Zhangzhou Geothermal Field.

For basins with small lateral extent, such as the Liaohe and Longyan Basins, we have shown that a simple analytic model in which "cold" or "hot" water flows in a confined horizontal aquifer can explain the observed exponential variation of temperature and heat flow along the direction of water flow. For basins of large extent and for geothermal fields, a more sophisticated numerical modeling is necessary. We have used finite difference and finite element methods to model the thermal regime of the Beijing Geothermal Field. The results agree well with the observations.

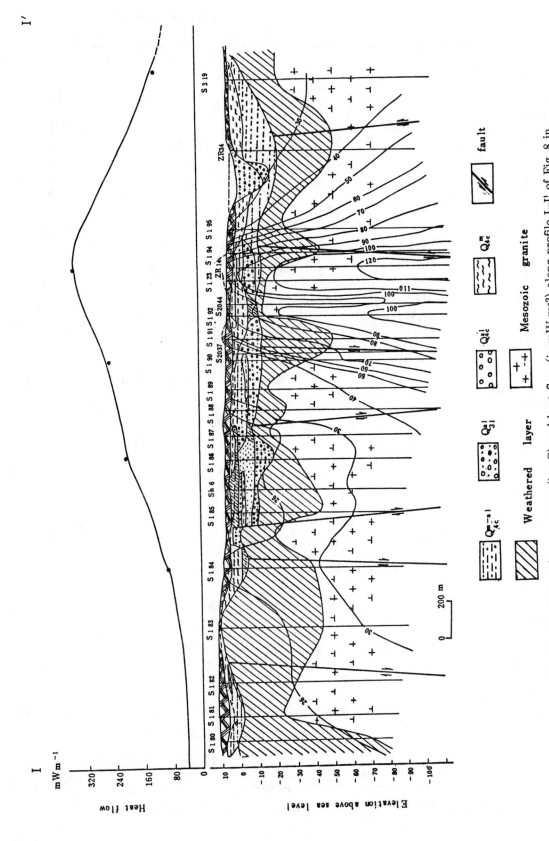

Fig. 10. Variation of temperature (in °C) and heat flow (in mW m⁻²) along profile I–I' of Fig. 8 in Zhangzhou Geothermal Field.

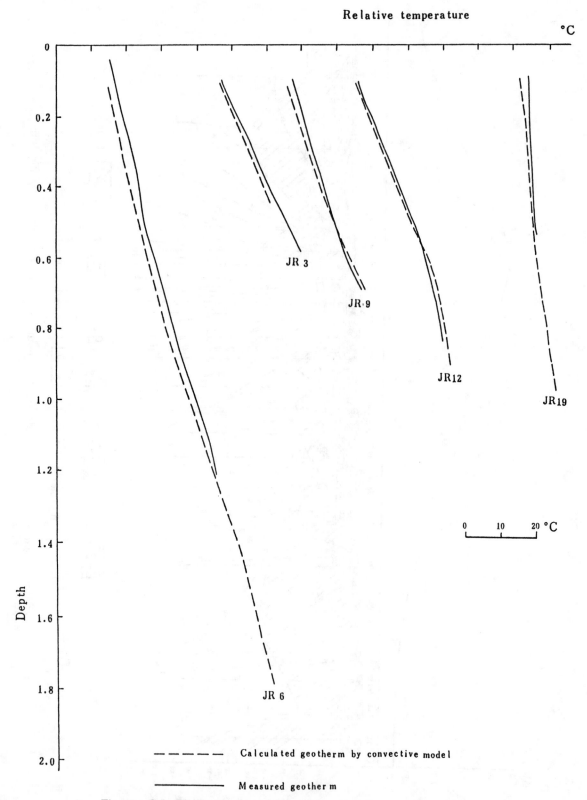

Fig. 11. Calculated and measured geotherms in Beijing Geothermal Field.

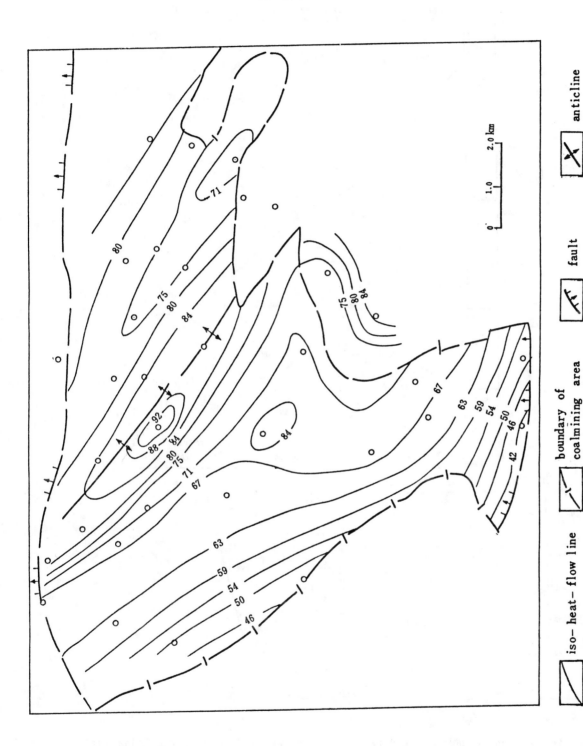

Fig. 12. Heat flow map (in mW m⁻²) for Xinzheng Basin of Henan Province, North China (Modified from Deng, 1988). o = heat flow measurement site.

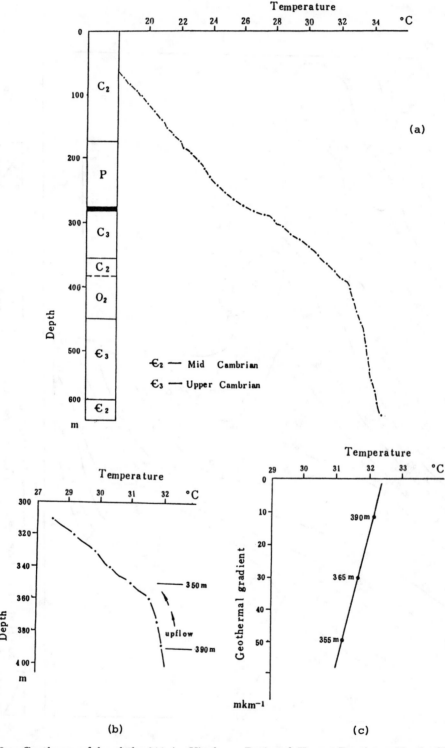

Fig. 13.   Geotherm of borehole 911 in Xinzheng Basin of Henan Province, North China.   (a) Temperature profile of the borehole, (b) Convex shape of the profile between 300 and 400 m indicating an upflow of thermal water, (c) Temperature vs. temperature gradient plot (from Deng, 1988). O_2 = Mid Ordovician, C_2 = Mid Carboniferous, C_3 = Upper Carboniferous,   P = Permian, — = coal seam.

Acknowledgement. The authors wish to express their sincere thanks to Prof. A.E. Beck for encouraging us to submit this paper and for many valuable comments.

References

Boyle, J.M., 1979. Determination of recharge rates using temperature–depth profiles in well. *Water Resour. Res.*, 15(6), 1616–1622.

Bredehoeft, J.D. and Papadopulos, I.S., 1965. Rates of vertical groundwater movement estimated from the earth's thermal profile. *Water Resour. Res.*, 1(2), 325–328.

Deng, X., 1988. Effect of vertical groundwater movement on thermal field and case history for Xinzheng Basin of Henan Prov., North China. *Scientia Geologica Sinica* (in press).

Domenico, Z.A. and Palciauskas, Y.V., 1973. Theoretical analysis of forced convective heat flow transfer in regional groundwater flow. *Bull. Am. Geol. Soc.*, 84(12), 3803–3814.

Goyal, K.P. and Kassoy, D.R., 1980. Fault zone controlled charging of a liquid–dominated geothermal reservoir. *J. Geophys. Res.*, 85(B4), 1867–1875.

Kilty, K. and Chapman, D.S., 1980. Convective heat transfer in selected geologic situations. *Ground Water*, 18(4), 386–394.

Kassoy, D.R. and Zebib, A., 1978. Convection fluid dynamics in a model of a fault zone in the Earth's crust, *J. Fluid Mech.*, 88, 769–792.

Mansure, A.J. and Reiter, M., 1979. A vertical groundwater movement correction. *J. Geophys. Res.*, 84(B7), 3490–3496.

Parsons, M.L., 1970. Groundwater thermal regime in a glacial complex. *Water Resour. Res.*, 6(6), 1701–1720.

Smith, L. and Chapman, D.S., 1983. On the thermal effects of groundwater flow: 1. Regional scale systems. *J. Geophys. Res.*, 88(B1), 593–608.

Sorey, M.L., 1971. Measurement of vertical groundwater velocity from temperature profiles in wells. *Water Resour. Res.*, 7(4), 963–970.

Stegena, L., 1982. Water migration influences on the geothermics of basins. *Tectonophysics*, 83, 91–99.

Torrance, K.E. and Chen, V.W., 1980. A model of hydrothermal convection in an aquifer. *J. Geophys. Res.*, 85(135), 2254–2258.

Tsang, C.F., Buscheck, T. and Doughty, C., 1981. Thermal energy storage: a numerical simulation of Auburn Univ. field experiments. *Water Resour. Res.*, 17(3), 647–658.

Wang, J.–y. and Wang, J.–a., 1986. Heat flow measurements in Liaohe Basin, North China, *Science Bull.*, 31(10), 686–689.

Wang, J.–y., Wang, J.–a., Wang, Y.–l., and Zhang, Z.–y., 1986. Terrestrial heat flow in Lower Liaohe Basin, North China. *Scientia Geologica Sinica, 1,* 16–29.

Xiong, L.–p. and Gao, W.–a., 1982. Characteristics of geotherm in uplift and depression. *Acta Geophys. Sinica,* 25(5), 448–456.

Xiong, L.–p. and Wang, J.–a., 1988. Temperature measurements in a small basin with deep–circulating thermal water, case history in Longyan basin of Fujian Prov., SE China. *Earth Sci.* (in press).

Xiong, L.–p. and Zhang, J.–m., 1988. Relationship between geothermal gradient and the structure form of basement rock in North China Basin. *Acta Geophys. Sinica* (in press).

Xiong, L.–p., Zhang, J.–m., and Sun, H.–w., 1985. Mathematical simulation of geotemperature and heat flow patterns, *J. Geodynamics, 4,* 45–61.

Zhang, J.–m. and Xiong, L.–p., 1986. *Application of Finite Element Method in Geothermics,* Science Press, Beijing, China, pp.138 (in Chinese).

THE CONTROVERSY OVER THE SIGNIFICANCE OF THE HYDRODYNAMIC EFFECT ON HEAT FLOW IN THE PRAIRIES BASIN

J. A. Majorowicz

Institute of Earth and Planetary Physics and Department of Physics, University of Alberta,
Edmonton, Canada, T6G 2J1

Extended Abstract. The horizontal and vertical fluid velocities inferred from the heat flow studies in the Prairies basin in Canada by various authors are usually one or two orders of magnitude larger than velocities reported by most hydrogeologists. This problem has surfaced in many recent publications by Bachu (1985, 1988), Bachu et al. (1987), Garven (1985, 1986, 1987) and Majorowicz, et al. (1987). For the most part hydrogeological studies are based on measurements of core permeabilities and well tests which provide information on scales smaller than the basinal scale. Estimates of the continuum parameter called permeability may not be well represented by laboratory measurements or short term borehole packer tests or aquifer tests (Kirally, 1975; Garven, 1986). Permeability of regional fractures may not be measurable in these scales and local porosity and permeability may not represent a basinal scale permeability.

There are two divergent points of view on the matter of the significance of regional scale hydrodynamic effect on heat flow in the Prairies basin in Canada.

One point of view is that the hydrodynamic effect for heat flow is too small to be accountable due to the small fluid velocities so that the heat is transferred by conduction only (Bachu, 1985; Bachu et al., 1987). The argument presented by Bachu (1985) and Bachu et al. (1987) comes from studies of areas in Alberta, Cold Lake and Swan Hills. Hydrogeological studies in these regions and dimensional analysis applied to heat transfer processes show that measured permeability, and indeed the fluid velocities, are too low to play a significant role in the transport of terrestrial heat in Alberta. Bachu (1988) attributes regional distribution of heat flow in Alberta characterized by low heat flow in the foothills of the Rocky Mountains and high heat flow in the lowlands in northeastern Alberta,

close to the Precambrian Shield to crustal thickening and/or radiogenic heat generation in the basin. In that, Bachu (1988) is side stepping the foundation of the interpretation of the geothermal field pattern given by Anglin and Beck (1965). The latter attributed the northward increase of geothermal gradient to crustal thickening which would provide a great amount of radioactive heat generation per column of unit area. However this is at odds with the recent results of the crustal thickness studies in Western Canada (see Kanasewich et al., 1987, Fig. 10) and basement heat generation in the Prairies basin by Jones and Majorowicz (1987a, Fig. 2).

Another point of view is that the regional heat flow variations have been profoundly influenced by the hydrodynamic effect. Heat flow observations seem to suggest this view. Majorowicz and Jessop (1981) and Hitchon (1984) noticed on the regional and local scales respectively that higher hydraulic head areas which are related to higher topographic elevations are associated with lower geothermal gradients and heat flow whereas higher geothermal gradients and higher heat flow areas are associated with lower elevations. That can not be attributed to refraction of heat as shown by numerical modelling for central Alberta (Majorowicz et al., 1987). The contribution of refraction is relatively small as shown by the model of Luheshi and Jackson (1986) for southern Alberta. The model results of Luheshi and Jackson (1986) show that the temperature field in southern Alberta is almost entirely dominated by gravity forced fluid convection effects above the Paleozoic. The statistical relationship between heat flow and topographic elevations exists for the heat flow estimates based on shallower Mesozoic formations rather than deeper Paleozoic strata (Majorowicz et al., 1985c; Majorowicz et al., 1986). This has been interpreted as being due to the effect of reduction of heat flow by downward crossformational fluid flows in high fluid potential recharge areas and heat flow advection due to upward crossformational flows in

Copyright 1989 by
International Union of Geodesy and Geophysics
and American Geophysical Union.

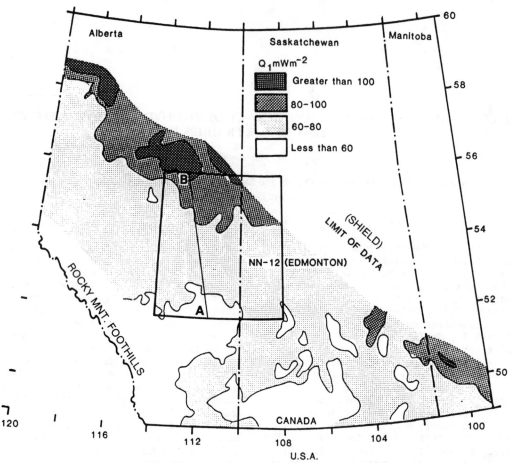

Fig. 1. The location of the profile A-B against the map of average terrestrial heat flow for Mesozoic and Cenozoic formations. The NN-12 map sheet area for which detailed heat flow studies has been done (Majorowicz et al. 1987) is indicated.

low fluid potential fluid discharge areas (Majorowicz and Jessop, 1981 and Majorowicz et al., 1985a, b).

Changes in heat flow with depth have been observed through comparison of the average values based on heat flow estimates for the sedimentary section above and below the Paleozoic erosional surface (Majorowicz et al., 1984; Majorowicz et al., 1985a). The dominant clastic lesser thermal conductive unit above that surface is underlain by the predominantly calcareous-evaporitic unit. It has been observed that in the large central part of the basin, between regional recharge areas and regional discharge areas, heat flow is fairly constant with depth and Q=60-80 mW m^{-2} (Majorowicz et al., 1985a, b). Heat flow variations with depth of the order of ±(20-40) mW m^{-2} have been reported in fluid recharge and discharge areas. Heat flow variations with depth have also been reported by Andrews-Speed et al. (1984) for the Western North Sea basin.

The existence of high heat flow (Q>80 mW m^{-2}) in the Prairies basin next to a low heat flow

shield characterized by an average heat flow of 44 ±7 mW m^{-2} (Drury, 1985) and 40 ±8 mW m^{-2} (Jessop and Lewis, 1978) for the Churchill and Superior provinces respectively is another argument for the existence of the advection component of heat flow in the basin. In the north (65°N – 70°N) heat flow generally increases from the shield and platforms in the east towards the foldbelt in the west, whereas in the southern parts of the basin, the heat flow increases toward the shield. This is attributed to hydrodynamically controlled forced convection in the southern portion of the area but such heat transport is prevented by continuous permafrost in the north (Majorowicz et al., 1988).

The lack of correlation between heat flow estimated in the sediments of the Prairies basin and the heat generation variation in the crystalline basement has been also observed (Majorowicz and Jessop, 1981; Jones and Majorowicz, 1987a) and has been used as an argument to support the hypothesis that most of the crustal and mantle heat which flows into the sediments is

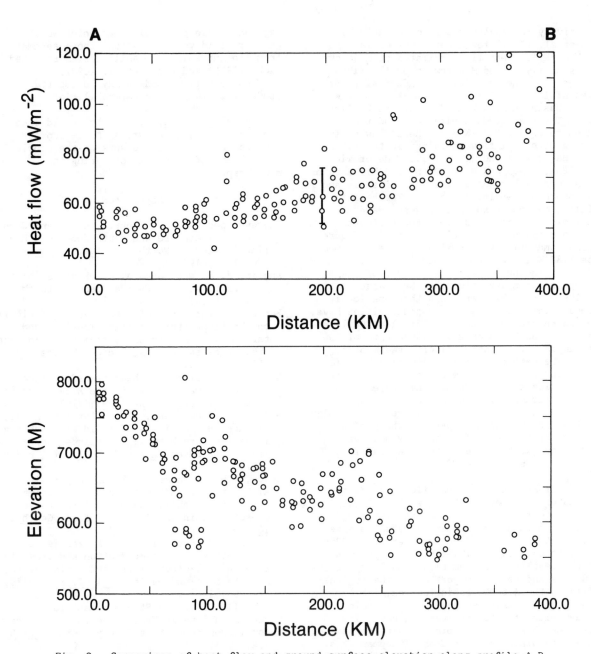

Fig. 2. Comparison of heat flow and ground surface elevation along profile A-B of Figure 1. Sample error bar for the heat flow data is shown.

redistributed by fluid motion through permeable strata (Majorowicz and Jessop, 1981).

One of the major problems in supporting the second point of view is the question of the accuracy of heat flow estimates. This depends on the accuracy of bottom hole temperature data and the accuracy of the thermal conductivity estimates based on net rock analysis and average rock conductivities. Average rock conductivities are based on measurements of thermal conductivities of rock samples from sediments. Even if the temperature gradients are accurate, heat flow estimates depend very much on the value of conductivity assumed for shales. There seems to be a general problem in thermal conductivities of shales which was critically examined by Blackwell and Steele (1981) and Sass and Galanis (1983). Blackwell and Steele (1981) estimated values for the conductivities of shales to be $1.05 - 1.25$ W m^{-1}K^{-1} which agree with the conductivities measured using the needle probe method by Sass and Galanis (1983). The average conductivity for shales reported recently by Beach et al. (1987) for Alberta shales is 1.38 W m⁻¹K⁻¹. Heat flow estimates of Majorowicz and Jessop (1981) and Majorowicz et al. (1986) were based on the much larger conductivity for shales of 1.5 W m⁻¹K⁻¹ - 1.9 W m⁻¹K⁻¹ for the Alberta basin and 1.5 W m⁻¹K⁻¹ for the Williston Basin. The heat conductivity for sandstone was assumed to be 4.2 W m⁻¹K⁻¹. A heat flow profile checking high heat flow anomalies in the Alberta basin (Majorowicz and Jessop, 1981; Majorowicz et al., 1985a; Jones and Majorowicz, 1987b) based on the assumed higher conductivities for shale and sandstone has been investigated. The heat flow profile location is shown in Figure 1 and the heat flow variations and topographic elevation variations against distance are shown in Figure 2. The conductivity of shales has been assumed to be 1.38 W m⁻¹ K⁻¹ and the conductivity of sandstone, another major rock builder of the Mesozoic sedimentary column has been assumed to be 3.1 W m⁻¹K⁻¹ which are less than previously assumed. These average rock conductivities have been chosen according to Beach et al. (1987) and they have been based on the measurements of 300 core samples from central Alberta sedimentary rocks. The methods used in estimating heat flow and a description of the error analysis is given by Majorowicz et al. (1987). The increase in heat flow accompanying decreasing topographic elevations can be observed in Figure 2. The high heat flow in the northern part of the profile confirms the high heat flow anomaly observed in the previous work although absolute heat flow values are less by up to 20% of the previous estimates.

The increase in paleo-heat flow towards the northeastern part of the basin with heat flow high next to the "cold" shield has been suggested by organic maturation geochemical studies (see, Beaumont et al., 1985; Majorowicz et al., 1985d; Majorowicz et al., 1987) which is in agreement with the results of 2D numerical modelling of the fluid flow and heat transport along the regional profile in the Alberta basin (Garven, 1987).

The proposal that regional basinal scale hydrodynamics affects the regional heat flow pattern has been made for other basins worldwide, such as the Uinta basin in Utah, (Chapman et al., 1984; Willet and Chapman, 1987), the Liaohe Basin in North China (Wang Ji-Yang et al., 1985), the Western North Sea basin (Andrews-Speed et al., 1984) and the Denver, Kennedy and Williston Basins of the Great Plains of the U.S.A. (Gosnold, 1985) and the Great Arthesian Basin in Australia (Cull and Conley, 1983).

Heat flow studies seem to indicate the existence of the hydrodynamic effect. They therefore suggest the existence of larger regional permeabilities than those inferred in the lab and from well tests. Fractures with spacing in the order of 100 to 1,000 m have been suggested by Bredehoeft et al. (1983) and Garven (1986) as being sufficient to explain larger

regional scale permeabilities which would allow greater fluid velocities than those estimated from lab and well permeabilities as reported by Bachu (1985) and Bachu et al. (1987).

Thus heat flow studies may be important to the study of basinal scale permeabilities.

Acknowledgements. The author would like to express his thanks to Dr. A. M. Jessop for giving him the opportunities to work on the problem, and Miles Ertman for help in preparing the manuscript.

References

Andrews-Speed, C. P., Oxburgh, E. R. and Cooper, B. A., Temperature and depth dependent heat flow in Western North Sea, A.A.P.G. Bull., 68, 1764-1781, 1984.

Anglin, F. M. and Beck, A. E., Regional heat flow pattern in Western Canada, Can. J. Earth Sci., 2, 176-182, 1965.

Bachu, S., Influence of lithology and fluid flow on the temperature distribution in a sedimentary basin: A case study from the Cold Lake area, Alberta, Canada, Tectonophysics, 120, 257-284, 1985.

Bachu, S., Analysis of heat transfer processes and geothermal pattern in the Alberta basin, Canada, J. Geophys. Res., 93, 7767-7781, 1988

Bachu, S., Sauveplane, C. M., Lytviak, A. T., Hitchon, B., Analysis of fluid and heat regimes in sedimentary basins: Techniques for use with large data bases, A.A.P.G. Bull., 71, 822-843, 1987.

Beach, R.D.W., Jones, F. W. and Majorowicz, J. A., Heat flow and heat generation estimates from the Churchill basement of the western Canadian basin in Alberta, Canada, Geothermics, 16, 1-16, 1987.

Beaumont, C., Boutilier, R., Mackenzie, A. S. and Rullkoetter, J., Isomerization and aromatization and paleothermometry and burial history of Alberta foreland basin, A. A. P. Bull., 69, 546-566, 1985.

Blackwell, D. D. and Steele, Heat flow determinations in Kansas and their implications for midcontinent heat flow pattern, EOS, 62, p. 392, 1981.

Bredehoeft, J.D.S., Neuzil, C. E. and Milly, P.C.D., Regional flow in the Dakota Aquifer: A study of the role of confining layers, U. S. Geol. Survey Water-Supply Paper 2237, p. 45, 1983.

Chapman, D. S., Keho, J. H., Bauer, M. S. and Picard, M. D., Heat flow in the Uinta basin determined from bottom hole temperature data, Geophysics, 49, 453-466, 1984.

Cull, J. P. and Conley, D., Geothermal gradients and heat flow in Australian sedimentary basins, B.M.R. Journal of Austral. Geol. Geophys., 8, 329-337, 1983.

Drury, M. J., Heat flow and heat generation in the Churchill province of the Canadian Shield and their paleotectonic significance, Tectonophysics, 115, 25-44, 1985.

Garven, G., The role of regional fluid flow in the genesis of the Pine Point deposits, Western Canada sedimentary basin, Economic Geology, 80, 307-324, 1985.

Garven, G., The role of regional fluid flow in the genesis of the Pine Point deposit, Western Canada sedimentary basin - a reply, Economic Geology, 81, 1015-1020, 1986.

Garven, G., Quantitative models for stratabound ore genesis in sedimentary basins, Proceedings of the third annual Canadian/American Conference on hydrogeology: "Hydrogeology of sedimentary basins", B. Hitchon, Ed., U. S. National Water Well Assoc., Dublin, Ohio, 69-74, 1987.

Gosnold, W. D., Heat flow and groundwater flow in the Great Plains of the United States, J. Geodyn., 4, 247-265, 1985.

Hitchon, B., Geothermal gradients, hydrodynamics and hydrocarbon occurrences, Alberta, Canada, A.A.P.G. Bull., 68, 713-743, 1984.

Jessop, A. M. and Lewis, T. J., Heat flow and heat generation in the Superior Province of the Canadian Shield, Tectonophysics, 50, 55-77, 1978.

Jones, F. W. and Majorowicz, J. A., Regional trends in radiogenic heat generation in the precambrian basement of the Western Canadian basin, Geophys. Res. Letter, 14, 268-271, 1987a.

Jones, F. W. and Majorowicz, J. A., Some aspects of the thermal regime and hydrodynamics of the Western Canadian sedimentary basins, Proceedings, Fluid Flow in Sedimentary Basins and Aquifers, Geological Society of London, June 12-13, 1985, Geological Society Special Publication 34, 79-85, 1987b.

Kanasewich, E. R., Hajnal, Z., Green, A. G., Cumming, G. L., Mereu, R. F., Clowes, R. M., Morel-A-L'Huissier, P., Chiu, S., Macrides, C. G., Shahriar, M., and Congram, A. M, Seismic studies of the crust under the Williston basin, Can. J. Earth Sci., 24, 2160-2171, 1987.

Kirally, L., Rapport sur l'état actuel des connaissances dans le domaine des caracteres physiques des roches kartiques: Internat. Union Geol. Sci., Ser. B, No. 3, 53-67, 1975.

Luheshi, M. N. and Jackson, D., Conductive and convective heat transfer in the sedimentary basins, Proceedings of the 1st IFP Exploration research conference, Carcans, France, 1985, Thermal modelling in sedimentary basins, J. Burrows (Ed.), Editions Technip, Paris, 219-234, 1986.

Majorowicz, J. A. and Jessop, A. M., Regional heat flow patterns in the Western Canadian Sedimentary basin, Tectonophysics, 74, 209-238, 1981.

Majorowicz, J. A., Jones, F. W. and Jessop, A. M., Geothermics of the Williston basin in Canada in relation to hydrodynamics and hydocarbon occurrences, Geophysics, 51, 767-779, 1986.

Majorowicz, J. A., Jones, F. W. and Jessop, A. M., Preliminary geothermics of the sedimentary basins in the Yukon and Northwest territories (60°N - 70°N) - estimates from petroleum bottom hole temperature data, Bull. Can. Petr. Geol., 36, 39-51, 1988.

Majorowicz, J. A., Jones, F. W., Lam, H. L. and Jessop, A. M., The variability of heat flow both regional and with depth in southern Alberta, Canada: Effect of groundwater flow? Tectonophysics, 106, 1-29, 1984.

Majorowicz, J. A., Jones, F. W., Lam, H. L. and Jessop, A. M., Regional Variations of heat flow differences with depth in Alberta, Canada, Geophys. J. R. astr. Soc., 81, 479-487, 1985a.

Majorowicz, J. A., Jones, F. W., Lam, H. L. and Jessop, A. M., Terrestrial heat flow and geothermal gradients in relation to hydrodynamics in the Alberta basin, Canada, J. Geophys., 4, 265-283, 1985b.

Majorowicz, J. A., Jones, F. W., Lam, H. L. and Nguyen, C. D., Topography and the subsurface temperature regime in the Western Canadian Sedimentary basin: Implications for low enthalpy geothermal energy recovery, Geothermics, Spec. issue, U. N. Seminar on Geothermal Energy, Florence, 1984, 14, 75-187, 1985c.

Majorowicz, J. A., Rahman, N., Jones, F. W. and McMillan, N. J., The paleogeothermal and present thermal regimes of the Alberta basin and their significance for petroleum occurrences, Bull. Can. Petr. Geol., 33, 12-21, 1985d.

Majorowicz, J. A., Jones, F. W., Ertman, M. E., Linville A., and Osadetz, K., Heat flow in the Edmonton-Cold Lake region of the Western Canadian Sedimentary basin and the influence of fluid flow, Prodeedings of the third annual Canadian/American Conference on Hydrogeology: "Hydrogeology of Sedimentary Basins", B. Hitchon (Ed.), U. S. National Water Well Assoc., Dublin, Ohio, 151-158, 1987.

Sass, J. A. and Galanis, S. P., Jr., Temperatures, thermal conductivity, and heat flow from a well in Pierre Shale near Hayes, South Dakota, U. S. Geol. Survey. Open File Report, 83-25, 1983.

Wang, Ji-Yang, Wang Ji-An, Xiang Liang-Ping and Zhang Tu Ming, Analysis of factors affecting heat flow density determinations in the Liaohe basin, North China, Tectonophysics, Spec, Issue, A. E. Beck (ed.), 121, 63-78, 1985.

Willet, S. D. and Chapman, D. S., On the use of thermal data to resolve and delineate hydrogeologic flow system in sedimentary basins: An example from the Uinta basin, Utah, Proceedings of the third annual Canadian/American Conference on hydrogeology: Hydrogeology of Sedimentary Basins", B. Hitchon (ed.), U. S. National Water Well Assoc., Dublin, Ohio, 159-168, 1987.

SIMULATION OF THE HYDROTHERMAL SYSTEM AT HIRAYU HOT SPRINGS, JAPAN

Kozo Yuhara, Yasuhiro Fujimitsu and Takashi Okabe

Faculty of Engineering, Kyushu University,
Hakozaki, Fukuoka 812, Japan

Abstract. A three-dimensional mathematical
model has been developed for Hirayu Hot Springs
in central Japan. In the model a steady-state
flow of the liquid phase was assumed. A rather
small area (1.1 km x 1.1 km) was chosen as the
model region and geological information,
including drilling data and the results of
magnetotelluric surveys, was used to determine
the underground structure of this region.
Temperature and pressure boundary conditions were
treated as parameters. With these preconditions
trial and error calculations were performed to
find the temperatures and flow velocities which
is the most suitable for the well logging
temperature data and the total discharge rate of
the hot springs. From this model the
distribution of underground water flows, checked
with a tracer test, was determined. Before this
work, it was thought that the inflow to the
region was mainly from the eastern side, where an
investigation tunnel is being bored. However,
this work shows that there is an additional
inflow of almost the same magnitude from the
southern side.

Introduction

Hirayu Hot Springs are located in a valley
that is 1200 m above sea level and are surrounded
by the Hida Mountains, the North Alps of Japan.
Mount Yake, the active volcano, is about 5 km
from the hot springs (Fig. 1). Route 158 passes
through the mountain area but is closed by snow
in winter time. Therefore, under the Abo Pass,
which is located at the border of Gifu and Nagano
Prefectures, a tunnel is being constructed. Much
hot and cold waters well out from the western
side of the investigation tunnel and it is
suspected that these losses may have some
influence on the nearby Hirayu Hot Springs.
Therefore geophysical and geochemical, as well as
geological and other, investigations are being

Copyright 1989 by
International Union of Geodesy and Geophysics
and American Geophysical Union.

carried out. Details of the data have been
reported by Chubu Regional Construction Bureau
[1986]. The purpose of this paper is to report
the results of a mathematical modeling of this
region which is used to infer the heat sources
and distribution of underground flows (direction,
quantity, etc.) of thermal waters.

There are many examples of modeling of heavily
convecting water systems. Models for opened
geothermal systems have been described by Yuhara
et al. [1979], Akibayashi et al. [1981],
Akibayashi et al. [1981a, 1981b] and Akibayashi
et al. [1982]. For simplicity, in most of these
models it was assumed that only the density of
water was affected by temperature. However, a
model presented by Pritchett [1981] allowed for a
temperature dependent viscosity. Hanaoka [1980]
used a finite element method to take account of
the topographical effect of thermal convection in
a mountainous region.

In the present paper, it is assumed that the
thermal waters in the Hirayu Hot Springs area are
always in the liquid phase and that the flow
regime in the permeable media is steady-state.
In addition to the effects of buoyancy and
topography, the temperature dependence of the
viscosity is also taken into account. The finite
difference method is used to determine the three-
dimensional flows in a rather small area (1.1 km
x 1.1 km).

Summary of Data Available in the Hirayu Hot Springs Area

Hirayu Hot Springs are surrounded by mountains
which peak at about 2000 m above sea level (Fig.
1). The geological structure of the region is
mainly composed of a basement rock of Paleozoic
chert, slate and schalstein, underlying
Quaternary volcanic products and alluvial
deposits. The basement rock is exposed in the
southern and western mountains, and Mount
Akandana is composed of andesite tuff breccia.
The Abo Marsh, about 2 km to the SE of the hot
springs, is thought to be a recharge area for the
hot springs in Hirayu. Several wells (see Fig.

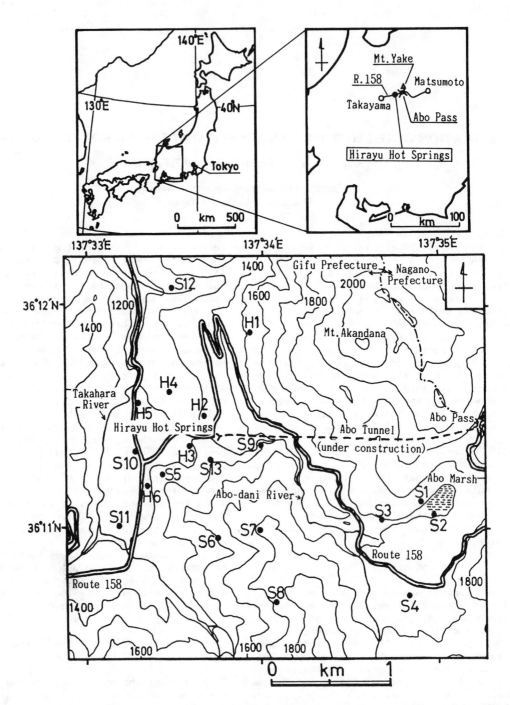

Fig. 1. Map of Hirayu Hot Springs. The investigation wells are indicated by black
spots. Elevation contour interval is 100 m.

1) have been drilled to obtain geological and
thermal data. Generally, the basement rock has
low permeability except in the vicinities of
wells S5, S9 and S13. In the formation of
volcanic products (mainly tuff breccia) there are
both low permeability (H2, H4) and high

permeability (H5) sections. The mean thermal
conductivity from H2-H5 wells of the basement
rock is 3.6 W m^{-1} K^{-1} but only 1.3–1.7 in the
other formations. A contour map of temperature
at the 1150 m level (see Fig. 2) based upon
thermal logging data indicates two high

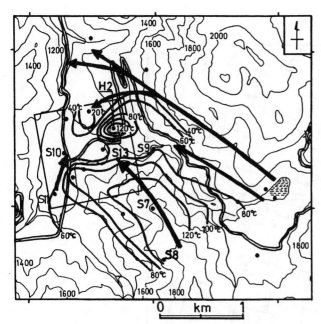

Fig. 2. Temperature contour map of 1150 m above sea level based on the temperature logging data, and inferred underground water flows by the tracer test [Chubu Regional Construction Bureau, 1986]. The square shows the modeling area.

temperature regions. One is in the vicinity of the H2 well and the other is near S7 (Fig. 2). The homogenization temperatures of the fluid inclusions in the cores of S7-S10 and S13 wells were measured and found to cover a wide range, 102-278 $^{\circ}$C, from wells S7,S8 and S13. Since the well heads are on the southern mountain, we assume that this region has been hot in the past and might still be a source of heat.

Tracer tests, using fluorescent dyes and activable tracers, were used to determine the flow direction of ground water in the region (see Fig. 2). Since some tracers from the Abo Marsh input point were detected at Hirayu Hot Springs after 13-49 days, there must be some routes where the underground water flows are faster than others.

A magnetotelluric survey [Mogi et al., 1986] indicates a low resistivity anomaly extending from S7 to H5 and two high resistivity anomalies, one in the vicinity of H2 and the other extending from S5 and H6 towards S6 (Fig. 3). The location of the low resistivity anomaly is in reasonable agreement with one of the underground water routes inferred from the tracer tests.

Analysis of tritium concentration in the waters suggests that the hot springs can be divided into four groups; A - in the northern part with the hot water having a high concentration of tritium with a value that is nearly equal to that of the water gushing into the investigation tunnel; B - in the eastern

part, the concentration is low; C - at the foot of the southern mountain, the value is a little lower than that of the Abo-dani River and D - in the upper reaches of the Takahara River where the value is nearly equal to that of the Abo-dani River.

Governing Equations

We assume that the thermal fluid is liquid and incompressible water flowing through a permeable medium. To construct a 3-D mathematical model we also assume that (1) the water flow is in a steady-state described by Darcy's Law, (2) the variation of water density can be described by the Boussinesq approximation, that is, density varies with temperature only in the gravity term of the equation of motion and (3) the viscosity of water varies inversely with temperature.

The governing equations are therefore

$$\mathrm{div}\ V = 0 \tag{1}$$

$$\rho_o\, C \nabla (\, T\, V\,) = \nabla (\, \lambda \nabla T\,) \tag{2}$$

and

$$V = -\, k\, (\nabla P + \rho^{*}\, g\,)\, /\, \mu \tag{3}$$

$$\rho^{*} = \rho_o\, [\, 1 - \beta\, (\, T - T_o\,)] \tag{4}$$

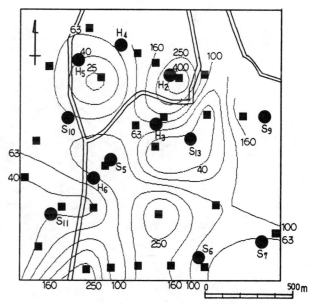

Fig. 3. 7.8 Hz scalar apparent resistivity map. Unit is ohm m. The geometric mean of the values measured in N-S and E-W directions of the induction coil are used [Mogi et al., 1986]. The investigation wells and the survey points are indicated by solid circles and solid squares, respectively.

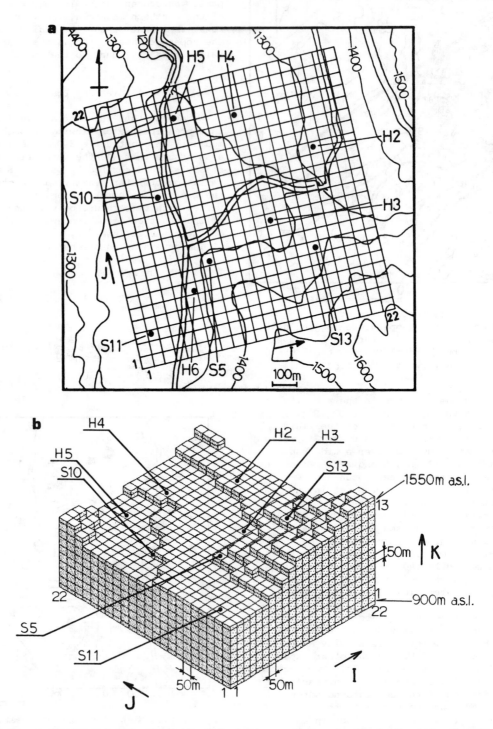

Fig. 4. a: Horizontal extent of the calculation region. b: Three-dimensional extent of the calculation region.

where V is the filter velocity vector, k the permeability of the aquifer, μ the dynamic viscosity of the thermal water, P the fluid pressure, ρ the density of the thermal water, ρ_o the density of the water at the reference temperature, g the gravitational acceleration, β the expansion coefficient, T the temperature in degree C, T_o the reference temperature, C the specific heat of the thermal water and λ the thermal conductivity of the porous medium saturated with thermal water.

The variation of viscosity with temperature utilizing the equation of the kinematic viscosity by Wooding [1957] is given by

$$\mu = 0.33 \; p^* / 10000 / T \qquad (5)$$

Substitution of equations (3), (4) and (5) into (1) and (2) results in two partial differential equations for pressure and temperature. For numerical calculations these partial differential equations are replaced with finite difference equations which are then expanded explicitly so that the iterative method of successive over-relaxation (SOR) [Young, 1954] can be used. After obtaining distributions of pressure and temperature by this method, the distribution of filter velocity components using equation (3) can be obtained.

Modeling

Model Region

The region chosen for modeling (Fig. 4) is essentially determined by the limited area of the magnetotelluric surveys and investigation wells. Of the wells indicated, H6 was too shallow (about 40 m deep) so the data from this well are not included in the analysis. The greatest depth, 900 m above sea level, of the region used for analysis was determined by the elevation of the bottom of the deepest investigation well. To reproduce the mountainous topography of the area the region was divided into cubes of 50 m sides; additional blocks 25 m thick are attached to the top and the bottom of the calculation region, as shown in Fig. 4b, and similar blocks were added to the vertical boundaries but are not shown in Fig. 4b.

Boundary Conditions

Fukutomi [1951] has produced a standardized table of month, latitude and altitude dependent temperatures at 1 m depth. Hirayu Hot Springs are located at 36 degrees 11 minutes north latitude, so we take the annual mean ground temperature at this area to be given by

$$T_s = 14.8 - 0.00358 \; H \qquad (6)$$

where T_s is the temperature in degree Celsius and

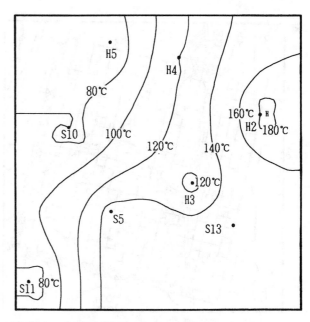

Fig. 5. Temperature distribution at the bottom boundary (875 m above sea level). Black spots show the investigation wells.

H is the altitude of the ground surface in meter above sea level.

The bottom temperature distribution was assigned as shown in Fig. 5 which is a solution for the two-dimensional thermal conduction equation in which the extrapolated temperatures from well logging data and homogenization temperatures of fluid inclusions data were fixed. The temperature distribution on the vertical boundaries was taken as a parameter and is discussed later.

At all boundaries except the bottom, pressure distributions were assigned; the bottom boundary was assumed to be impermeable. The underground water levels in the investigation wells were used to obtain the pressure distribution on the top boundary, by having an empirical least squares relationship between the altitude of any subsurface point, the ground surface altitude immediately above that point and the pressure at that point. In addition, the region was divided into three units of different relation (as shown in Fig. 6a) for each of which there was a coefficient for the altitude relational equations as shown in equations (7), (8) and (9)

$$P = 44.6066 + 0.0697 \; Z_s - 0.10593 \; Z_D \quad \text{(for A)} \quad (7)$$

$$P = 7.6873 + 0.0750 \; Z_s - 0.08008 \; Z_D \quad \text{(for H)} \quad (8)$$

$$P = 35.0599 + 0.0393 \; Z_s - 0.06728 \; Z_D \quad \text{(for S)} \quad (9)$$

where Z_D is the altitude of any point in meter above sea level, Z_s the altitude of the ground surface above that point in meter and P the

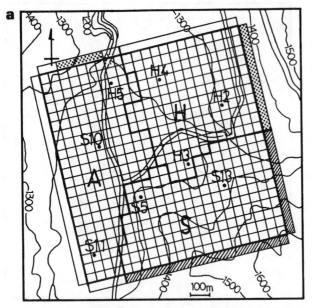

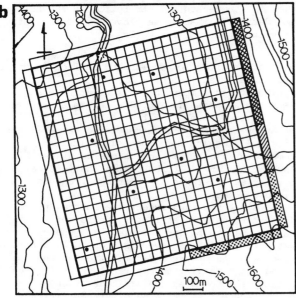

Fig. 6. a: Boundary condition of pressure. The ground surface boundary is divided into three parts by permeability. The vertical boundaries where the pressure is taken as a parameter are indicated by dotted and hatched bands. b: Vertical boundary condition of temperature. The parts where temperature is taken as a parameter are indicated by dotted, hatched and meshed bands.

pressure at that point in bars. Pressure values of all boundaries were calculated using the above equations and the pressures of the top of the region were fixed as boundary conditions. Initial pressures at the vertical boundaries were

also assigned by these equations but varied later as a parameter.

Physical Properties of Thermal Water

The physical properties used for thermal water are shown in Table 1. The expansion coefficient

TABLE 1. Physical properties of thermal water

Density at the reference temperature	1045.95 kg m^{-3}
Reference temperature	0.0 °C
Expansion coefficient	9.374x10^{-4} 1 K^{-1}
Specific heat	4190.0 J kg^{-1} K^{-1}
Gravitational acceleration	9.80665 m s^{-2}

and density of thermal water at its reference temperature, necessary for equation (4), were obtained by a least squares fit to standard data from 15 to 300 °C [Tokyo Astronomical Observatory ed., 1981].

Fitting Process

To construct a suitable model two basic steps are necessary. In the first, the approximate distribution of thermal conductivity and permeability is determined in the region of interest, and in the second the temperatures and rates of recharge from outside the area of interest are adjusted until the model fits (i) the observed distribution of hot water discharge, (ii) the flow pattern obtained by the tracer test and (iii) the temperature profiles of the eight investigation wells.

At the beginning of the first step the region of interest was divided into five segments on the basis of the apparent resistivity map (Fig. 3) and the lithological sections inferred from eight investigation wells. Permeability values were assigned to each segment on the basis of data obtained from the well tests, and conductivity values were assigned on the basis of measurement made on cores; in each segment the physical properties were assumed to be isotropic. Pressures on vertical boundaries were fixed at values obtained from equations (7)-(9) and temperatures on vertical boundaries were kept at the values assigned by linear interpolation between temperatures at the ground surface and the bottom boundaries. The extent and the values of permeability and thermal conductivity for each segment were gradually adjusted in order to obtain the most suitable underground structure. In the second step, hydrostatic pressure of a few hundred meters were added partially to allow for possible potential flow caused by the topography, and solutions of the two-dimensional thermal

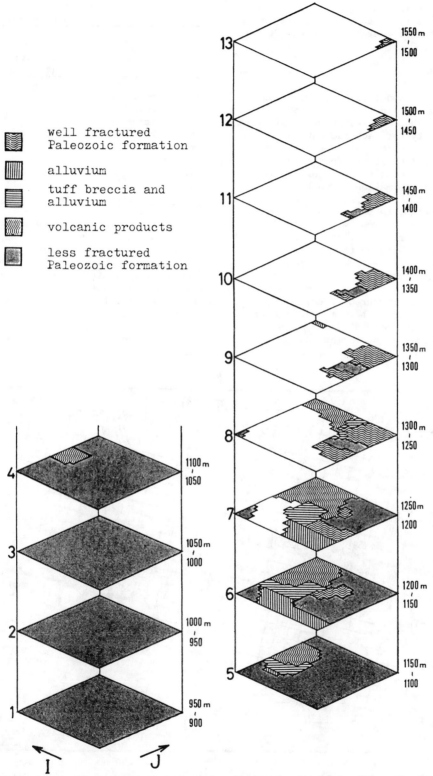

well fractured
Paleozoic formation

alluvium

tuff breccia and
alluvium

volcanic products

less fractured
Paleozoic formation

Fig. 7. Reproduced underground structure for the most suitable model. The layer number and the altitudes (above sea level) of the top and the bottom of each layer are shown.

conduction equation were assigned to the vertical boundaries from the first to the sixth layer so as to reproduce high temperatures in the deeper parts of the region.

Most Suitable Model

The most suitable model (Fig. 7) was obtained after 26 modifications in the first step and 6 in the second step. The values of permeability and thermal conductivity for each segment in the region of interest are shown in Table 2. The higher apparent resistivity to the southern mountain (Fig. 3) matches the part of the less fractured Paleozoic formation shown from the first to the 11th layers in Fig. 7. The magnetotelluric method is very sensitive to the lower resistivity layer [Mogi et al., 1987]. Therefore, it is inferred that the higher resistivity layer extends to the deep at the place where indicates the higher apparent resistivity. On the other hand, the lower apparent resistivity from S7 to S13 matches the portions of the well fractured Paleozoic formation shown from the 6th to the 13th layers. The higher resistivity near H2 matches the relatively low permeability formation of volcanic material between the 4th and 9th layers.

TABLE 2.   Permeability and thermal conductivity for each segment

	Permeability $(10^{-15} \, m^2)$	Thermal conductivity $(W \, m^{-1} \, K^{-1})$
Less fractured Paleozoic formation	20	3.6
Well fractured Paleozoic formation	100	3.6
Volcanic products	150	1.7
Tuff breccia and alluvium	1500	1.5
Alluvium	1700	1.5

Figure 8 compares the measured (a) and calculated (b) hot water discharges. In our model self flowing discharge was assumed although there are some wells where hot water is being pumped up in addition to the artesian activity in Hirayu Hot Springs. Total hot water discharge of this model (96 l s^{-1}) is less than that of the measured value of 134, but for most blocks in which there is only artesian activity, there is good agreement between the measured and calculated values.

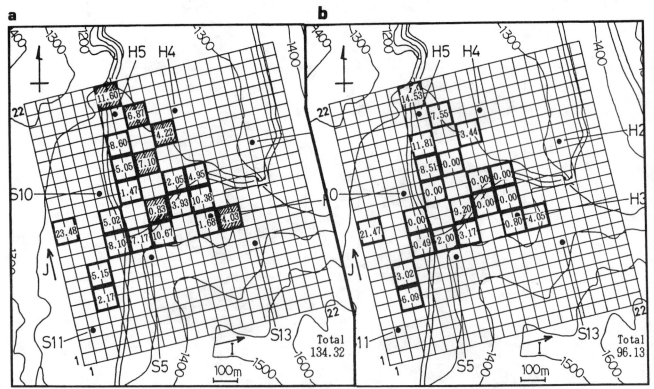

Fig. 8.   a: Measured values of hot water discharge from the hot springs [Chubu Regional Construction Bureau, 1986].   Unit is l s^{-1}.   Hatched blocks include artesian wells only.   b: Calculated values of hot water discharge at the blocks corresponding to those of Fig. 8a.

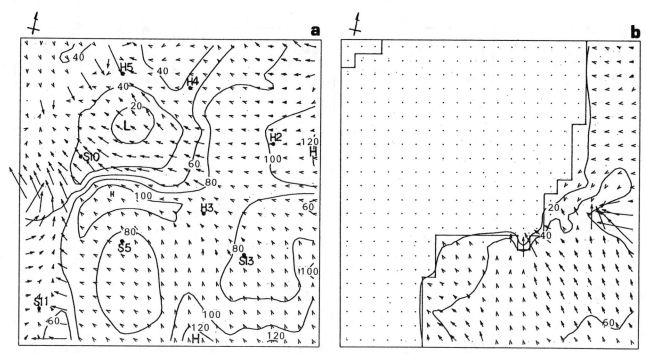

Fig. 9. a: Calculated horizontal flow pattern and temperature distribution in the sixth layer (1150-1200 m above sea level). The longest arrow corresponds to the filter velocity of 71.9 x 10^{-7} m s^{-1}. b: Calculated horizontal flow pattern and temperature distribution in the eighth layer (1250-1300 m above sea level). The longest arrow corresponds to 77.3 x 10^{-7} m s^{-1}. Unit of temperature is $^{\circ}$C.

The similarity between the measured and calculated temperatures at 1150 m altitude may be seen by comparing Figures 2 and 9a. In both figures two high temperature zones (shown by H in Fig. 9a) and one low temperature zone are found. The vertical profiles of the measured and calculated temperatures in the eight investigation wells are compared in Fig. 10. Quantitative agreement is reasonable for S11 and H2 and qualitatively reasonable for H3, H5 and S5, but poor for H4, S10 and S13.

The vertical boundaries where pressures and temperatures are taken as parameters are shown in Figures 6a and 6b. In Fig. 6a, on the boundary shown by dotted marks, a hydrostatic pressure equivalent of 50 m was added to the value obtained by using equations (7) and (8), while on the hatched boundary, hydrostatic pressure equal to 250 m depth was added. In Fig. 6b, on the boundary shown by dotted marks, the value of 140 $^{\circ}$C was assigned from the first layer to the 7th layer. On the hatched boundary, linearly interpolated values between the top and the bottom temperatures were applied from the first layer to the 6th layer, and on the meshed boundary, a few degrees Celsius was added to the values obtained by the second step from the first layer to the 7th layer.

Discharge and recharge rates through each boundary are shown in Table 3. The difference between discharge and recharge rates is 0.19 l s^{-1} that indicates good convergence of the

TABLE 3. Discharge and recharge rates through each boundary

	Top	North	South	East	West	Sum
Discharge	499.67	72.02	15.09	11.89	25.83	624.50
Recharge	116.88	37.83	187.99	204.78	77.21	624.69

Unit is 10^{-3} m^{3}s^{-1}.

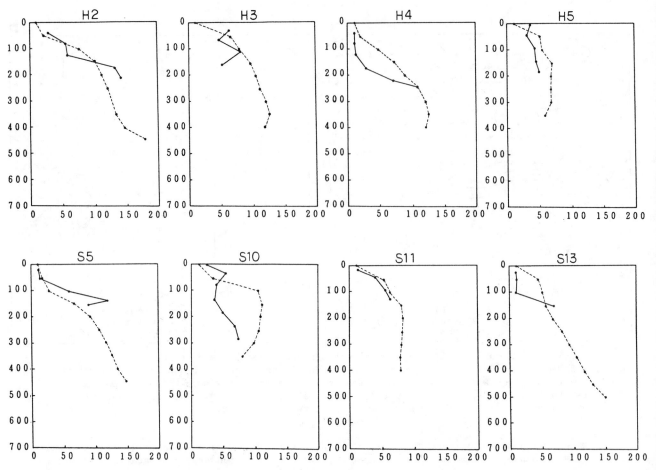

Fig. 10. Vertical temperature profiles of eight investigation wells. Horizontal axis represents temperature (°C) and vertical axis represents the depth from well head (m). Solid and broken lines indicate measured and calculated values, respectively.

calculation for this model. The data indicate two main areas of recharge where the rates are nearly equal, one on the eastern and the other on the southern boundary.

In Fig. 9a the flows from the northeast, southeast and south inferred from the tracer test are shown. Horizontal flow pattern in the 6th layer (1150-1200 m above sea level) seem to represent a typical flow pattern for the area. In addition to these a flow from the northwest is indicated. The flow from the southeast is clearer in the 8th layer (1250-1300 m above sea level) as shown in Fig. 9b. The vertical flow pattern for the north-south section is shown in Fig. 11 and that for the east-west section in Fig. 12. In Fig. 11 two main upflow regions are found in the area of the hot springs and a downflow occures between them. Fig. 12 includes well H2 that indicates high temperature. There is also a downflow region in the middle part of this section where there is a river. The eastern

part of this downflow suggests a recharge from the river.

Discussion

In a field situation, we are unlikely to obtain a unique interpretation because there may be a lack of data from key areas and errors in the data. Therefore, we stress that although our preferred model fits the field data well, a further search might yield other models in reasonable agreement with the field data.

In Fig. 2, it appears that there is a localized upflow of hot water in the vicinity of H2. However, the local nature may be an artifact of the limited data set because there are no investigation wells to the northeast of H2 and it is possible that the high temperature zone is broader than indicated on Fig. 2. There is some evidence to support this possibility, that is, the investigation tunnel bored 3.5 km to the east

S-N

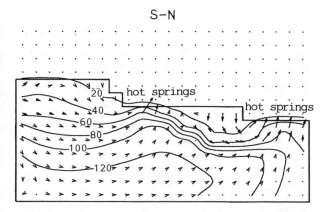

Fig. 11. Calculated vertical flow pattern and temperature distribution (°C) at I=9 (N-S section, see Fig. 4a). The longest arrow corresponds to the filter velocity of 41.3×10^{-7} m s^{-1}.

of H2 passes through hot dry rocks. Our interpretation of the data from the H2 well is that H2 is not in the region of maximum temperature, that there is therefore no upflow in this region, and that thermal waters flow from the east as indicated in Fig. 12.

In the model, there are four main flows from the northeast, southeast, south and northwest with recharges from eastern and southern boundaries surpassing those across the other boundaries. Although the main purpose of the present work was to construct a reasonable model of Hirayu Hot Springs, an additional objective was to estimate the influence of the construction of the tunnel on the future behavior of the hot springs. Because of the large recharge on the eastern boundary, the construction of the tunnel will cause a decrease in the discharge rate of the eastern and northern parts of the hot

W-E

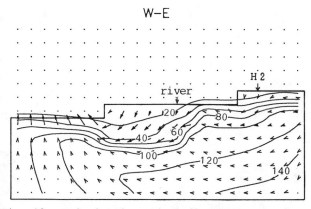

Fig. 12. Calculated vertical flow pattern and temperature distribution (°C) at J=15 (E-W section, see Fig. 4a). The longest arrow corresponds to the filter velocity of 29.2×10^{-7} m s^{-1}.

springs. In particular, the water temperature in the northern part may fall because of mixing with cold underground water from the northwest. However, in the western and southern parts, especially in the vicinity of S5 and S11, there may be little change because the recharge area is supposed to be the southern mountain.

It is usual to insert a transition region between the region to be analyzed and the outer boundaries so as to reduce the influence of oversimplified boundary conditions in a real field situation where the boundary conditions are essentially unknown. Nevertheless, in the model there are no buffer regions because setting of such regions causes a reduction of the effect of mountainous topography. Consequently, the values near the vertical boundaries may be influenced by the boundary conditions.

Conclusion

In spite of the mountainous topography and a large number of variables that have to be considered, a reasonable model of the Hirayu Hot Springs area can be constructed. The resultant model has a complicated flow pattern and includes two dominant recharge flows with nearly equal recharge rates (about 195 +/- 5 l s^{-1}) from the eastern and southern boundaries. The model obtained here is based on the investigations carried out in this area and will be a base model for a parametric study of the influence of the construction of the Abo Tunnel on the hot springs.

Acknowledgments. The geological, geophysical and geothermal data used in the present paper were supplied by Chubu Regional Construction Bureau, Ministry of Construction.

References

Akibayashi, S., Yuhara, K. and Nakanishi, S., A study on the natural convection in the opened geothermal reservoir(II) - an experiment on the steady-state natural convection in a porous medium with a permeable top surface -, J. Geothermal Res. Soc. Japan, 2, 107-119, 1981(in Japanese with English abstract).

Akibayashi, S., Yuhara, K., Nakanishi, S. and Tanaka, S., A study on the natural convection in the opened geothermal reservoir(III) - comparison of the experimental and the theoretical results -, J. Geothermal Res. Soc. Japan, 3, 15-30, 1981a(in Japanese with English abstract).

Akibayashi, S., Yuhara, K., Nakanishi, S. and Tanaka, S., A study on the natural convection in the opened geothermal reservoir(IV) - effects of the length of impermeable portion at the top surface -, J. Geothermal Res. Soc. Japan, 3, 101-114, 1981b(in Japanese with English abstract).

Akibayashi, S., Yuhara, K. and Tanaka, S., A
    study on the natural convection in the opened
    geothermal reservoir(V) - application of the
    opened thermal convection model to the
    practical geothermal system -, J. Geothermal
    Res. Soc. Japan, 4, 143-158, 1982(in Japanese
    with English abstract).
Chubu Regional Construction Bureau, Ministry of
    Construction, Technical Report - Abo Tunnel -,
    1986(in Japanese).
Fukutomi, T., On the possibility of the
    prospection of hot-springs by the geographical
    distribution of underground temperature in 1 m
    depth (the 1st report), Geophys. Bull. Hokkaido
    Univ., 1, 1-14, 1951(in Japanese with English
    abstract).
Hanaoka, N., Numerical Model experiment of
    hydrothermal system - topographic effects -,
    Bull. Geol. Surv. Japan, 31, 321-332, 1980.
Mogi, T., Ehara, S., Yuhara, K., Asoshina, K. and
    Fujimitsu, Y., A comparison of the ELF-MT
    resistivity structure and the thermal structure
    in Takenoyu geothermal area, J. Geothermal Res.

Soc. Japan, 9, 237-249, 1987(in Japanese with
    English abstract).
Mogi, T., Yuhara, K. and Kajiwara, T.,
    Geoelectrical structure of Hirayu Hot Springs
    by the ELF-MT method, Annual Meeting Geothermal
    Res. Soc. Japan Abstracts, 58, 1986(in Japanese).
Pritchett, J. W., The LIGHTS code, System,
    Science and Software -R-80-4195-R1, 1981.
Tokyo Astronomical Observatory ed., Rika nenpyo
    (Chronological Scientific Tables), Maruzen Co.,
    Ltd., 1981(in Japanese).
Wooding, R., Steady state free thermal convection
    of liquid in a saturated permeable medium, J.
    Fluid Mech., 2, 273-285, 1957.
Young, D., Iterative methods for solving partial
    differential equations of elliptic type, Trans.
    Amer. Math. Soc., 76, 92-111, 1954.
Yuhara, K., Akibayashi, S. and Nakanishi, S., A
    study on the natural convection in the opened
    geothermal reservoir(I) - relation between the
    mode of convection and the size of heat source
    -, J. Geothermal Res. Soc. Japan, 1, 69-83,
    1979(in Japanese with English abstract).

EFFECTS OF PERMEABILITY ANISOTROPY AND THROW ON THE TRANSMISSIVITY IN THE VICINITY OF A FAULT

D. Bernard , M. Danis , M. Quintard

L. E. P. T. - E.N.S.A.M. (UA C.N.R.S. 873),
Esplanade des Arts et Métiers - 33405 Talence Cedex France,
C.R.P.G. 15, rue N.-D. des Pauvres, BP 20, 54501 Vandoeuvre Cedex France

Abstract. The effect of a fault on the flow in an aquifer was quantified numerically. The simple situations considered were characterized only by the throw of the fault and the permeability anisotropy of the aquifer.

Using these results one can take into account the influence of the fault in two dimensional (2-D) hydrological models simply by introducing a singular hydraulic head loss on the fault boundary. Two numerical formulations for the singularity were tested and the results compared to a reference solution obtained from a three dimensional (3-D) finite element model.

From these results, it is possible to obtain additional information on the hydrology of real aquifers as shown in the case of a Lower Triassic sandstone aquifer in the Vittel area (Vosges, France) for which we solve the inverse problem in order to predict the local rock permeabilities.

Introduction

Over the past twenty years numerical simulation of aquifer systems has developed greatly. The starting point for such hydrological studies is the choice between a flow model which represents the aquifer as a porous medium or as a connected network of fractures (Bear, 1972). In early work, aquifers were represented as homogeneous isotropic porous media with hydrological properties estimated from pumping tests (see review by Kruseman and de Ridder, 1970); however later work assumed they were fractured media (see for instance, Louis,

1969; Gale, 1982; Long et al, 1982; Andersson et al, 1984; de Marsily, 1985). Thus, models have evolved from the homogeneous isotropic porous medium concept to that of the multilayer aquifer type, and then to the concept of distributed heterogeneous media (Rojaz et al, 1984).

In this paper we consider the effect of a fault on the flow in an aquifer which can be represented as a porous medium. The fault affects the apparent transmissivity of the aquifer in two different ways:

1 - The throw produces a distortion of the streamlines near the fault leading to an increase in the flow head loss. This distortion depends on both the fault throw amplitude and the vertical permeability, i.e., the anisotropy of the aquifer.

2 - The petrophysical properties can be changed in the vicinity of the fault. If present, this effect produces a change in the head loss which can be important. However, predictive quantification is difficult, and in this paper we assume this effect is negligible.

Normally the selection of transmissivity coefficients is achieved by successive numerical trials without using precise rules, although it is recognized in the literature that the effective transmissivity reduction depends strongly on the fault throw (Daly et al, 1980; Faye and Prowell, 1982). In our study methods to calculate the hydraulic head field are tested. This begins with the numerical calculation of the additional head loss due to the fault for the simple case of a homogeneous anisotropic aquifer with constant thickness and vertical fault throw. Then this additional head loss is introduced in classical 2-D numerical hydrological models in two different ways:

Copyright 1989 by
International Union of Geodesy and Geophysics
and American Geophysical Union.

1 - the fault is treated as a mathematical boundary with a boundary condition for the head taking into account the additional head loss,

2 - the transmissivity is modified according to our results when the control volume associated with the numerical scheme is crossed by the fault.

The results obtained with these two different methods are compared to the reference solution obtained from a 3-D finite element model. A case study of a real aquifer shows that the transmissivities obtained through the trial procedure would have been evaluated more rapidly by our method.

Head Loss Across a Fault

Mathematical Model

We limit our study to a vertical 2-D aquifer section of constant thickness E (Fig. 1). The fault is vertical and the fault throw R is equal to R' multiplied by E. The length of the domain studied is L. In addition, we assume that although the porous medium is homogeneous, it can be anisotropic with respect to principal horizontal and vertical hydraulic conductivities, K_x and K_z. Such an aquifer corresponds to a large number found in sedimentary basins. We consider large values of L so that the velocities at the entrance and the exit of the domain can be assumed to be strictly horizontal and equal to V_o.

Groundwater flow satisfies the continuity equation (the density of the fluid is assumed to be constant):

$$\nabla \cdot V(x,z) = 0 \qquad (1)$$

where the water flux is given by Darcy's law (Bear, 1972; Greenkorn, 1983):

$$V(x,z) = - K \cdot \nabla h(x,z) \qquad (2)$$

where x and z are the horizontal and vertical coordinates, h is the hydraulic head, V is the Darcy velocity with components V_x and V_z; K is the hydraulic conductivity tensor where $K = \rho g k / \mu$ where ρ is the density of water, g the gravity acceleration, k the

intrinsic permeability tensor, and μ the dynamic viscosity of water. We have assumed that the principal directions of permeability are aligned with the cartesian coordinates such that

$$K = \begin{bmatrix} K_x & 0 \\ 0 & K_z \end{bmatrix} \qquad (3)$$

The boundary conditions are:

$$V_x = V_o \qquad \text{and} \quad V_z = 0 \qquad (4)$$

at the entrance and the exit of the aquifer, and

$$V \cdot n = 0 \qquad (5)$$

on the upper and lower impervious boundaries.

The governing flow equations can be rewritten by defining the following dimensionless variables:

$$x' = x / [E(K_x/K_z)^{1/2}]$$

$$z' = z / E$$

$$h' = h K_x / [V_o E (K_x/K_z)^{1/2}]$$

$$V_x' = V_x / V_o$$

$$V_z' = V_z / V_o \qquad (6)$$

The continuity equation (1) becomes:

$$\frac{\partial V_x'}{\partial x'} + \left(\frac{K_x}{K_z}\right)^{1/2} \frac{\partial V_z'}{\partial z'} = 0 \qquad (7)$$

Darcy's equation is written:

$$V_x' = - \frac{\partial h'}{\partial x'}$$

$$V_z' = - \left(\frac{K_z}{K_x}\right)^{1/2} \frac{\partial h'}{\partial z'} \qquad (8)$$

Combining these two equations leads to

$$\nabla'^2 h'(x',z') = 0 \qquad (9)$$

where ∇'^2 is the Laplacian operator related to x' and y'.

The new boundary conditions are:

$$V_x'=1 \quad \text{or} \quad \frac{\partial h'}{\partial x'}= -1 \; ; \quad x = 0, L' \qquad (10)$$

on the entrance and exit boundaries, and

$$V' \cdot n' = 0 \quad \text{or} \quad \nabla'h' \cdot n' = 0 \qquad (11)$$

on the impervious boundaries.

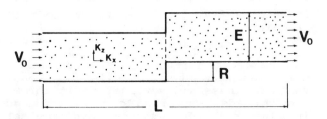

Fig. 1. Geometry

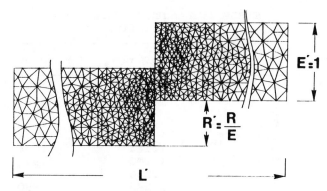

Fig. 2.   Finite Element Grid

The dimensionless boundary value problem does not depend on the anisotropy ratio K_z/K_x. If the fault is not vertical the anisotropy ratio appears in the calculation of the fault orientation angle in the dimensionless space (x',z').

Numerical Results

The boundary value problem defined by Eqs. (9), (10), and (11) is solved using the finite element model MODULEF, (Bernadou et al., 1985). An example of the discretization used is shown in Fig. 2. The head is approximated in each triangle by a piecewise continuous second order polynomial. The calculated head field corresponding to the geometry of Fig. 2 is shown in Fig. 3.

For each configuration, defined only by the dimensionless variable R', we are interested in two aspects:
1 - the quantification of the additional head loss due to the constriction created by the fault,
2 - the length of the zone influenced by the fault where the head gradient differs significantly from its uniform flow value.

Head Loss

The spatial variation of h' within the middle planes of the aquifer (trace noticed AA' in Fig. 3) is drawn in Fig. 4. Two distinct zones should be considered:
1 - at a sufficient distance from the fault, the gradient of h' is that of a uniform flow with the following value:

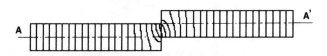

Fig. 3.   Hydraulic Head Field

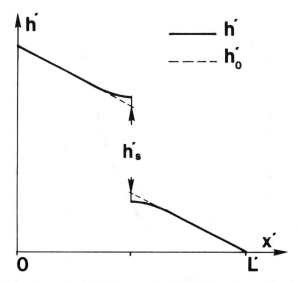

Fig. 4.   Definition of the Singular Head Loss h'_s

$$\frac{\partial h'}{\partial x'} = -1 \qquad (12)$$

2 - on both sides of the fault the gradient of h' differs from its uniform flow value; this defines the zone of influence of the fault.

The head along AA' can be approximated by the following function $h'_o(x')$:

$$x' < \frac{L'}{2} \qquad h'_o(x') = L' - x' + h'_s$$
$$\qquad (13)$$
$$x' > \frac{L'}{2} \qquad h'_o(x') = L' - x'$$

where h'_s is the additional hydraulic head loss due to the fault. The variations of h'_s with the relative throw R' are represented in Fig. 5.

Outside of the zone of influence $h'=h'_o$, but inside this zone $h' \neq h'_o$.

The zone of influence can be determined by the study of the function $h'-h'_o$. An example of this function map is shown in Fig. 6. It can be seen that the uniform flow is essentially disturbed near the boundaries on either side of the fault. If we introduce the distance from the fault L_ϵ such as:

$$| h' - h'_o | < \epsilon \qquad (14)$$

then each value of R' is associated with a value of L_ϵ, as indicated in Table 1 for $\epsilon=0.01$ and $\epsilon=0.001$.

The dimensional form of the additional hydraulic head loss h_s can be obtained from Eqs. (6)

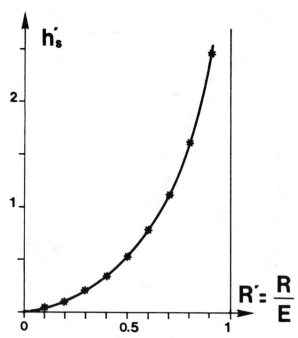

Fig. 5.   Singular Head Loss versus Relative Throw

TABLE 1. Length of the Fault's
Zone of Influence

Relative Throw	$L_{0.01}$	$L_{0.001}$
0.	.000	0.000
0.1	.606	1.330
0.2	.830	1.565
0.3	.974	1.680
0.4	1.084	1.798
0.5	1.144	1.894
0.6	1.202	1.946
0.7	1.250	1.984
0.8	1.295	1.990
0.9	1.313	2.030

transmissivity tensor, which is the product of $\underset{\sim}{K}$ by E in the case of a confined aquifer.

For our purposes, the idea is to introduce transmissivity anomalies in the 2-D hydrological model in order to approximate the real solution outside the zone of influence. These anomalies are introduced in the 2-D numerical model by two different methods:

1 - the fault is treated as a real boundary with a step change in the head on it,
2 - the transmissivity is locally modified in the vicinity of the fault.

The results obtained by the two methods are compared, for a test problem, to a reference solution given by a 3-D finite element model solving Eqs. (1)-(5).

The test problem consists of three successive segments representing a fault. The geometry and finite element grid are shown in Fig. 7. The boundary conditions are: impervious upper and lower boundaries, uniform flow at the entrance and exit of the aquifer domain. We will compare the 2-D field h to that computed with the 3-D program in the middle planes AA'.

First, it is assumed that the fault can be approximated by a set of segments on which

$$h_s = h_s' \ \frac{E}{K_x} \ (K_x/K_z)^{1/2} \ \underset{\sim}{V}_0 \cdot \underset{\sim}{e}_x \qquad (15)$$

where h_s' is the function of R/E represented in Fig. 5.

Calculation of the Hydraulic Head Field in an Aquifer with a Fault

The numerical simulation of groundwater flow through an aquifer is more often achieved in terms of a 2-D boundary value problem. For areal problems it is assumed that the hydraulic head only depends on the horizontal coordinates x and y, therefore it is possible to integrate Eq. (2) over the aquifer thickness to provide the following equation:

$$\underset{\sim}{Q}(x,y) = - \ \underset{\approx}{T} \cdot \underset{\sim}{\nabla} h(x,y) \qquad (16)$$

In this equation $\underset{\sim}{Q}$ is the flow-rate for the overall aquifer thickness and $\underset{\approx}{T}$ is the

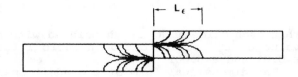

Fig. 6.   Fault's Zone of Influence

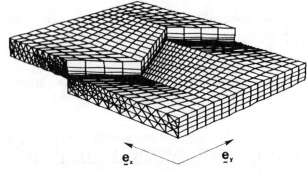

Fig. 7.   Geometry and Finite Element Grid of the 3-D Aquifer (R' = 0.8)

considering the simplification involved in the 2-D simulation.

The fault can also be represented by a transmissivity reduction. Our grid is chosen as to line up the fault with two rows of nodes as shown in Fig. 9(a). The transmissivity reduction that must be associated with the elements crossed by the fault can now be evaluated.

At a given point on the fault we consider a vertical section of the aquifer perpendicular to the fault and assume that in this section the velocity component perpendicular to the fault behaves as in the situation described earlier (Fig. 1). For simplicity the y-axis is taken to be

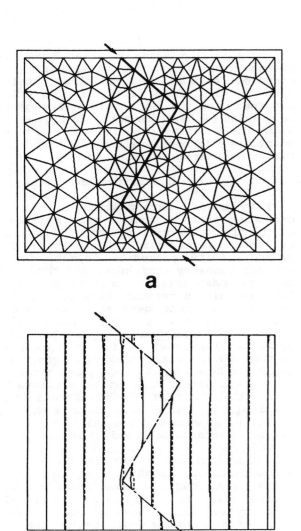

Fig. 8. Two-dimensional Flow Model with Internal Head Loss Across Fault ($R' = 0.5$): (a) Numerical Grid, (b) Hydraulic Head Results

relation (15) is valid. Thus, the fault is considered as a boundary with the following boundary condition:

$$h_{upstream} - h_{downstream} = h_s \quad (17)$$

where h_s is given by Eq. (15) in which $E\,V_{xo}$ is replaced by the component of Q_o normal to the strike of the fault.

The 2-D geometry and the associated finite element grid are shown in Fig. 8(a). The results are plotted (solid line) in Fig. 8(b) together with the 3-D reference solution (dashed line). The agreement between these two values is reasonable

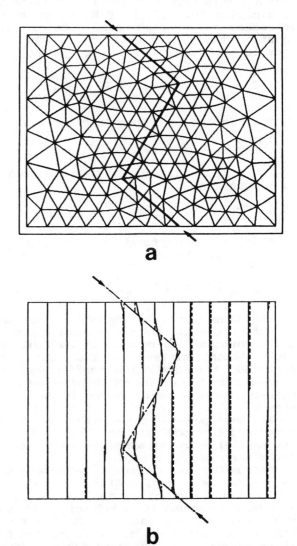

Fig. 9. Two-dimensional Flow Model with Transmissivity Reduction Across Fault ($R' = 0.5$): (a) Numerical Grid, (b) Hydraulic Head Results

parallel to the fault and, in order to approximate the true head loss, we look for a new transmissivity T_{fx} of the element crossed by the fault. The first order approximation of Eq. (16) is written

$$Q_{xo} = - T_{fx}(\Delta h / \Delta x) \qquad (18)$$

where Q_{xo} is the flow-rate normal to the fault, T_{fx} the modified transmissivity associated with the mesh, Δx the mesh size, and Δh the hydraulic head loss between the two points whose distance is Δx. The head loss Δh is the sum of a regular head loss δh due to the homogeneous transmissivity T_x within the element plus the additional head loss h_s. This latter value is calculated by multiplying Eq. (15) by E/E, in order to introduce on the right-hand side of this equation the characteristics of the 2-D hydrological model, i.e., $T_x = E K_x$ and $Q_{xo} = E V_{xo}$. We obtain:

$$h_s = [h'_s E (K_x / K_z)^{1/2} Q_{xo}] / T_x \qquad (19)$$

The regular head loss is given by:

$$\delta h = - (Q_{xo} \Delta x) / T_x \qquad (20)$$

Eq. (18) is transformed into the following equation:

$$Q_{xo} = - T_{fx}[(\delta h - h_s) / \Delta x] \qquad (21)$$

Then, by combining and simplifying Eqs. (18)-(21) we obtain:

$$T_{fx} = T_x / [1 + h'_s E (K_x / K_z)^{1/2} / \Delta x] = T_x / C \qquad (22)$$

The term C can be thought of as the reduction coefficient one should apply to the transmissivity normal to the fault in each element crossed by the fault. As expected, this reduction coefficient increases with increasing thickness E and mesh size Δx, and decreasing hydraulic conductivity K_z.

In the general case with any fault orientation we will use Eq. (22) with x as the spatial coordinate corresponding to the fault's normal direction.

This method is interesting because it can be generalized for discretization which do not take into account the fault geometry in an accurate manner. It is sufficient to look for the elements crossed by the fault and then to calculate new transmissivities for these elements according to the procedure described above. This is simpler numerically than the method with a step change in head on the fault boundary.

The agreement between the 2-D and 3-D models is fair in this case (Fig. 9(b)) but

not as good as the one obtained for a step change model (Fig. 8(b)). Actually, the computed fault related head loss obtained with the 3-D model is 0.363. The values obtained with the 2-D models are respectively 0.355 (2.2% difference) in Fig. 8, and 0.383 (5.5% difference) in Fig. 9. At this point, we can conclude that the method of Fig. 8 is more precise.

Field Application

The general method presented in the preceding section permits the calculation of the transmissivity reduction due to the fault for a given aquifer or reservoir. This transmissivity reduction depends on the aquifer geometry (thickness and throw) and on the permeability anisotropy ratio. However, the direct application of this method to a field case may lead to some difficulties, mainly because the detailed lithology of groundwater exploration boreholes is not known with sufficient precision as they are not core sampled. In general, E and R are known but K_x and K_z are generally poorly evaluated and must be estimated.

In this section we illustrate these difficulties in a real case study. In addition, we show how the use of our method allows a better understanding of the aquifer hydrology.

A numerical simulation was carried out on a Lower Triassic sandstone aquifer in the Vittel area, Vosges, France, (Fig. 10). This aquifer consists mainly of sandstone, with metric sized alternances of clay or marl. Groudwater recharge dominates in the southern part of the area where the aquifer

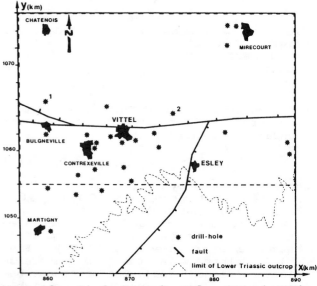

Fig. 10.   Study Area (x and y are given in the French Lambert coordinates)

outcrops. Regional flow is to the N-NE, towards the water supply boreholes and the centre of the Paris Basin.

Two major faults are distinguishable: the Esley fault, striking N-S, and the major Vittel fault, which is included in the 2-D hydrological model as a straight line, striking E-W, near the 1063 km N Lambert coordinate.

Flow was simulated in the area defined by the dashed line on Fig. 10. Boundary conditions were chosen according to the hydraulic balance of this area and the measured head, i.e., impervious boundaries, for the East and West boundaries and known hydraulic head on the South and North boundaries. The aquifer geometry (top and bottom) was mapped from the available borehole data. As the aquifer consists mainly of sandstone, it was assumed that the horizontal permeability was a function of the depth of burial, with correlations are taken from the literature (Redmond, 1962; Danis and Royer, 1986). The value of the aquifer transmissivity was deduced from the hydraulic conductivity and aquifer thickness outside the fault. In the area close to the Vittel fault a transmissivity reduction coefficient (Eq. 22) was adjusted by successive numerical trials.

Because the Esley fault behaves like an impervious boundary, the area of interest is limited, with x varying from 857 km to 876 km. Some of the available data describing

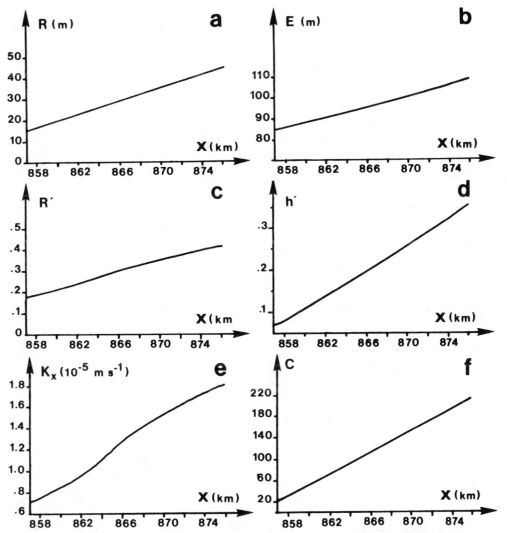

Fig. 11. Data: (a) Throw, (b) Thickness, (c) Relative throw, (d) Adimensional singular head loss, (e) Horizontal hydraulic conductivity, (f) Transmissivity reduction coefficient

the aquifer near the Vittel fault are shown in Figs. 11(a) through 11(f). The following points are emphasized:

a - The fault throw R, estimated from boreholes in the area, is assumed to vary linearly with distance, from 15m at point 857 km E (Lambert coordinate), to 45m at 876 km E (Crampon et al, 1975; BRGM, 1984).

b - The aquifer thickness E is assumed to be the same on either side of the fault; it varies almost linearly with the longitude (from 85m to 108m).

c - The relative throw R' is calculated at each point of the fault according to R'=R/E.

d - The dimensionless head loss h'_s is evaluated with the help of Fig. 5, and from the values of R' previously calculated.

e - The value of the horizontal hydraulic conductivity of the aquifer varies between 0.73×10^{-5} and 1.82×10^{-5} m s^{-1}.

f - The transmissivity reduction coefficient C, corresponding to a square grid ($\Delta x = 1000$m), varies from 20 in the far West to 210 at the Eastern boundary.

The lithology estimated from recent boreholes situated near the fault (boreholes n° 1 and 2) leads us to believe that the clay or marl proportion is constant all along the fault and equal to about 15% of the aquifer thickness.

All these data, as well as the detailed description of the methodology, are given in the paper by Danis and Royer (1986).

Interpretation: Fault Throw Effect

The direct calculation of the reduction coefficient C requires the knowledge of E, R', K_x and the permeability anisotropy ratio K_x/K_z. Unfortunately, this latter value is unknown. However, as the reduction coefficient has been adjusted by a trial-and-error technique, it is possible to solve a type of inverse problem which allows us to calculate the permeability anisotropy ratio. This provides additional information on the aquifer properties as shown in this section.

As the reduction coefficient C is known [Fig. 11(f)], as well as the aquifer thickness E [Fig. 11(b)] and R' [Fig. 11(c)], we can calculate the anisotropy ratio by using Eq. (22); we then obtain:

$$\frac{K_x}{K_z} = \left[(C-1) \frac{\Delta x}{h'_s E} \right]^2 \qquad (23)$$

where h'_s is the dimensionless head loss plotted in Fig. 11(d) and Δx is equal to 1000m.

It should be noted that the use of Eq. (22) is correct only if the porous medium behaves as a homogeneous medium for all the problems studied earlier. This is valid if the characteristic lengths of the heterogeneities (namely the strata thicknesses) are much smaller than the fault throw. For more details on the effect of local heterogeneities and the associated length scale constraints see Quintard and Whitaker (1987).

The calculated values of K_x/K_z as functions of the longitude are shown in Fig. 12(a). With these values and the corresponding values of K_x [Fig. 12(e)] we deduce the effective vertical hydraulic conductivity K_z. [Fig. 12(b)].

We see that the anisotropy ratio lies in a limited range between $1.02 \times 10^{+7}$ and $3.33 \times 10^{+7}$. This is obtained even if the reduction coefficient, C, which appears squared in Eq. (23), changes from 20 to 210. The anomalous small values at the West end (x < 863 km, shaded in Fig. 12) indicate, in our opinion, that this area is a shattered zone with petrological modifications related to the doubling of the fault (see Fig. 10).

While these results are interesting, it is possible to proceed even further and give reasonable estimates of the local permeabilities inside the aquifer as shown below.

Assuming that the layers of the aquifer consists of 85% homogeneous isotropic strata of sandstone, and 10% of homogeneous isotropic strata of clay or marl (Terrien et al., 1984) we can calculate the effective permeabilities K_x and K_z by using the following formulas:

$$K_x = 0.15 \, K_{clay} + 0.85 \, K_{sandstone}$$
$$K_z = 1/(0.15/K_{clay} + 0.85/K_{sandstone}) \qquad (24)$$

Using these equations and the previously calculated values of K_x and K_z we can determine the sandstone and clay permeabilities [Figs. 12(c)-12(d)].

For the Vosges sandstone, which are of the same area and stratigraphic level as the rocks of the Vittel aquifer, de Marsily (1981) gives hydraulic conductivity values of the order of 10^{-5} m s^{-1}, and for the clays, values in the range 10^{-9} to 10^{-13} m s^{-1}. The sandstone hydraulic conductivity estimated by our method varies steadily from 1.01×10^{-5} to 1.98×10^{-5} m s^{-1}, while the clay hydraulic conductivity increases from 0.52×10^{-13} to 1.26×10^{-13} m s^{-1}, in good agreement with the values indicated by de Marsily; the slight spatial variations could be linked to the aquifer burial which

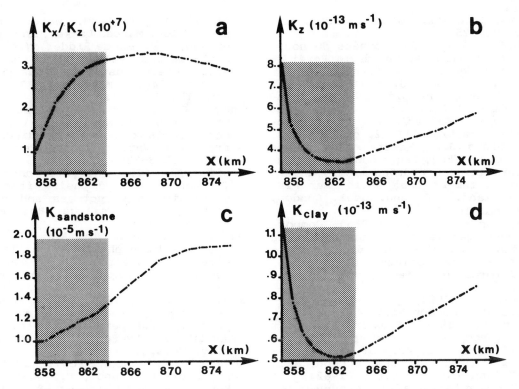

Fig. 12. Results: (a) Permeability ratio, (b) Vertical hydraulic conductivity, (c) Sandstone hydraulic conductivity, (d) Clay hydraulic conductivity

varies from 250m to 100m. The estimated hydraulic conductivities for the argillaceous rocks are also in agreement with other published data (see review by Brace, 1980).

We think that these results concerning sandstone and clay permeabilities and the anisotropy ratio are satisfactory, and show how our approach can contribute to the knowledge of an aquifer in the vicinity of a fault; no additional assumptions are needed (such as structural effects near the fault).

Conclusion

The effect of a fault on the flow in an aquifer is quantified numerically for simple situations characterized by fault throw, aquifer thickness and permeability anisotropy ratio.

These results permit a consideration of the fault influence in a 2-D hydrological model by adding a step-like hydraulic head loss at the crossing of the fault.

Although a direct application to a field problem will be difficult if no detailed data are available, we have shown that an inverse approach, illustrated on a field example, can provide additional information on the hydrological properties in the vicinity of the fault.

For other domains, such as petroleum engineering problems when more data are usually available, a direct application is probably easier. This, however, is beyond the scope of this study.

Acknowledgment. F. Lecocq and professor M. Pierre are gratefully acknowledged for their contribution. This work was supported in part by the "Institut National des Sciences de l'Univers" (ATP Transferts 1579).

References

Andersson, J., Shapiro, A.M. and Bear, J., (1984), A stochastic model of a fractured rock conditioned by measured information, Water Resources Res., vol. 20, 1, pp. 79-88.

Bear, J., (1972), Dynamics of fluids in porous media, Elsevier, New York.

Bernadou, M., et al., (1985), MODULEF: une bibliothèque Modulaire d'Eléments Finis, INRIA (INRIA-Rocquencourt, BP 105, 78153 Le Chesnay, France).

Brace, W.F., (1980), Permeability of crystalline and argillaceous rocks, Int. J.

Rock Mech. Min. Sci. & Geomech. Abst, vol. 17, pp. 241-251.

BRGM, (1984), Banque des données du sous-sol, SGR Lorraine, Avenue de la Forêt de Haye, 54500 Vandoeuvre, France.

Crampon, N., Le Roux, J. and Minoux, G., (1975), La nappe des grès du Trias Inférieur dans la région de Vittel et Contrexéville, Bulletin du BRGM, (deuxième série), section III, 2, pp. 119-128.

Daly, D., Lloyd, J.W., Misstear, B.D.R. and Daly, E.D., (1980), Fault control of groundwater flow and hydrochemistry in the aquifer system of the Castlecomer Plateau, Ireland. Q.J. eng. Geol. London, Vol.13, pp. 167-175.

Danis, M. and Royer J.J., (1986), Comportement hydrogéologique et thermique de la nappe des grès du Trias Inférieur dans le secteur de Vittel, Sciences de la Terre, Série Informatique, Nancy, vol. 25, pp. 33-56.

de Marsily, G., (1981), Hydrogéologie quantitative, Ed. Masson Paris, 215p.

de Marsily, G., (1985), Flow and transport in fractured rocks: connectivity and scale effects, Int. Ass. of Hydrogeologists, 17th Int. Congress, Tucson, AZ.

Faye, R.E. and Prowell, D.C., (1982), Coastal plain faults affect ground-water flow. Proceedings speciality conference environmentally ground water and soil management, USA, pp.251-252.

Gale, J.E., (1982), Assessing the permeability characteristics of fractured rock, Geological Society of America, Spec. pap. 189, pp. 163-181.

Greenkorn, R.A., (1983), Flow phenomena in porous media, Fundamentals and Applications in petroleum, water, and food production. Marcel Dekker, Inc. New York.

Kruseman, G.P. and de Ridder, N.A., (1970), Analysis and evaluation of pumping test data, Intern. Institute for Land Reclamation and Improvement, bull. 11, Wageningen, The Netherlands.

Long, J.C.S., Remer, J.S., Wilson, C.R. and Witherspoon, P.A., (1982), Porous media equivalents for networks of discontinuous fractures, Water Resources Res., vol. 18, 3, pp. 1253-1265.

Louis, C., (1969), Flows phenomena in jointed media and their effect on the stability of structures and slopes in rock, Imperial College, Rock Mechanics Progress Report.

Quintard, M. and Whitaker, S., (1987), Ecoulement monophasique en milieu poreux: effet des hétérogénéités locales, J. Méca. Théo. et Appliquée, vol. 6, 5, pp. 691-726.

Redmond, J.C., (1962), Effect of simulated overburden pressure on the resistivity, porosity and permeability of selected sandstones, The Pensylvania State University, Thesis.

Rojaz, J., Coleno, B., Fouillac, C., Gable, R., Giot, D., Iundt, F., Ouzounian, G., Menjoz, A. and Watremez, P., (1984), Le réservoir géothermique du Dogger en Région Parisienne. Exploitation Gestion. Hydrogéologie, Géologie de l'Ingénieur, vol. 1, pp. 57-85.

Terrien, M., Bergues, J., Sarda, J.-P. and Chaye d'Albissin, M., (1984), Etude expérimentale de l'anisotropie d'un grés et d'un marbre, Rev. I.F.P., vol. 39, 6, pp. 707-715.

FLUID FLOW IN CRYSTALLINE CRUST: DETECTING FRACTURES BY TEMPERATURE LOGS

Malcolm J. Drury

Geological Survey of Canada, 1 Observatory Crescent, Ottawa,
Ontario, K1A 0Y3 Canada

Abstract. The movement of fluids in low-permeability crystalline rock can cause significant perturbations to the normal conductive thermal gradient. The perturbations for different types of flow regime - flow between fractures, flow along fractures and flow into dead-end fractures - produce characteristic thermal anomalies that can be detected by standard techniques of temperature logging, and correlated with other geophysical logs such as single point resistance, porosity, (neutron-neutron) and density (gamma-gamma). The results of temperature logging of several closely-spaced boreholes in the granitic Lac du Bonnet batholith of the Canadian Shield are analysed in terms of the occurrence of fractures and the associated fluid flow regimes. Characteristic types of thermal anomaly are associated with fractures in zones defined by other geophysical and hydrological techniques, which implies that borehole temperature logging, in conjunction with measurement of thermal conductivity of core material, is a viable technique for the detection and mapping of fracture systems in crystalline rock. If the thermal properties of the rock penetrated by the borehole are known it is possible to quantify the rates of fluid flow. Thus, for example, combination of temperature logs and measurement of thermal conductivity of borehole core samples suggests the presence in the Lac du Bonnet batholith of a fracture zone down which water is flowing at the rate of approximately $0.3 \text{ gs}^{-1}\text{s}^{-1}$.

Introduction

The role of fluids in the formation and evolution of the earth's crust is of fundamental importance. Fluids are associated with metamorphism, with accumulation of hydrocarbons in sedimentary basins, with deposition of ores, and with the transfer of heat. The detection of large flows of hot brine at depths to greater than 9 km in the deep borehole of the Kola Peninsula [Kozlovsky, 1984] has shown that fluid-filled fracture zones can exist at considerable depths in the crust. There is mounting evidence for extensive fluid circulation, both in the past and at present, deep into the crust [e.g. Fyfe, 1986; Kerrich, 1986].

Heat flow and fluid flow are usually inextricably linked, and there is a large literature on the topic, mainly concerned with sedimentary basins. To those studying terrestrial heat flow, fluid flow in basins is a complicating factor, as it produces severe perturbations to the purely conductive regime [e.g. Majorowicz and Jessop, 1981; Stegena, 1982]. On the other hand, variations in heat flow related to the presence of fluids have the potential for being a valuable tool in the exploration for geothermal resources [e.g. Gale and Downing, 1986] and hydrocarbons [e.g. Majorowicz et al., 1986].

While the importance of fluids in controlling thermal regimes in sedimentary environments is well known, their thermal effects in low-permeability crystalline crust are perhaps no less significant. There, too, moving fluids can perturb the conductive thermal regime [e.g. Lewis and Beck, 1977, Hoisch, 1987]. As in sedimentary environments, the thermal perturbations are a hindrance to heat flow measurements, but they can be exploited. In this paper, one such exploitation is discussed. The thermal signatures of discrete fractures and fracture zones are briefly reviewed. A set of thermal data from a small area in a pluton of the Canadian Shield are then used to show how, in conjunction with other geophysical and geological techniques, they can be used to map fracture systems. The data presented are all from boreholes in the Lac du Bonnet batholith in the Canadian Shield, site of an Underground Research Laboratory (URL) operated by Atomic Energy of Canada Ltd. The batholith is a massive, homogeneous body of granitic - granodioritic composition in the Superior tectonic province of the shield. It has been the subject of intensive geophysical, geological and hydrological studies aimed at ascertaining the feasibility of underground disposal of hazardous materials in such stable bodies. Geothermal studies in the southern part of the batholith have been reported by Drury and Lewis [1983]. The data for the present paper are from a location approximately 15 km to the north of their site, centered on approximately 50° 15'N, 95° 52'W. Borehole temperature data were obtained at 3 m intervals with a portable logging system described by Drury and Lewis [1983]. Near surface data were omitted owing to the potential perturbations such as diurnal variations and other possible disturbances [e.g. Gatenby, 1977]. Two types of borehole were logged: fully cored,

Copyright 1989 by
International Union of Geodesy and Geophysics
and American Geophysical Union.

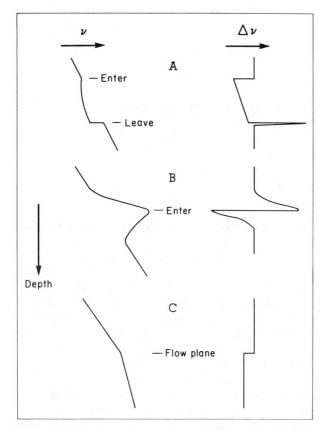

Fig. 1. Schematic representation of characteristic anomalies produced by fracture types A - C (see text), shown both as temperature (v) against depth and first difference filtered temperature (Δv). A is for downhole flow and C is for flow up a dipping fracture. Sense of each anomaly is reversed for reverse flow directions.

50 mm diameter holes drilled primarily for geophysical and geological research (URL series), and uncored, 250 mm diameter holes, used primarily for hydrological research (M series).

Thermal Indications of Fluid Flow in Discrete Fractures

Types of Thermal Anomaly

The movement of water in low-permeability rock takes place along those parts of discrete fractures or fracture systems that are open to flow. A substantial part of the total fracture porosity of crystalline rock may be in the form of isolated areas, or dead-end areas connected to flow paths but not allowing flow within them [e.g. Rasmuson and Neretnieks, 1986].

In a fractured rock body, water flow along and between fractures produces characteristic thermal anomalies that can be modelled quantitatively [e.g. Ramey, 1962; Drury et al., 1984; Beck and Shen, 1985]. Drury et al. [1984] considered and described examples of three types of channel for fluid flow:

A. that provided by the borehole itself (i.e. flow within the borehole), allowing flow between two fractures or aquifers that were previously unconnected but between which there is a difference of hydraulic potential;

B. isolated fractures that, when penetrated by a borehole, accept drilling fluid, thereby producing a transient 'spike' thermal anomaly that decays with time in a characteristic way [Drury and Jessop, 1982];

C. fracture systems that allow fluid flow up or down their dip (i.e. flow along a plane intersected by the borehole).

The characteristic appearance of the thermal anomalies associated with each of these fracture types is shown schematically in Fig. 1, as both temperature (v) and differential temperature (Δv) plots. It is assumed, in constructing these characteristic anomaly forms, that the undisturbed temperature increases uniformly with depth. There are two important points to note. Firstly, the simple forms of the anomalies shown here may be modified by changes of thermal conductivity across the fractures causing them. While types A and B should be unambiguously detectable by their characteristic Δv styles, a simple change of conductivity in the vertical direction would produce an identical form of temperature variation as type C. Hence it is essential that thermal conductivity be measured. Drury and Lewis [1983] interpreted temperature and conductivity data as indicating an example of a dipping fracture zone along which water was flowing (type C) in a borehole in the southern part of the Lac du Bonnet batholith. In that case the anomaly occurred at a known dipping fracture zone, penetrated by the borehole. Lewis and Beck [1977] inferred the presence of such a dipping zone under an area of approximately 5 km^2 in the Canadian Shield from the variation of heat flow measurements, made in 71 boreholes.

Secondly, the anomalies shown are for discrete fractures, whereas closely-spaced fractures may occur in zones up to several metres wide. In such zones, the distinct style of the discrete thermal anomalies is likely to be smoothed. In the following discussions, reference to a "fracture" includes such zones. Furthermore, an individual fracture or fracture zone may permit more than one of these types of flow. Hence the thermal signatures of fracture zones may be complex.

The transient 'spike' anomaly associated with a fracture that has accepted drilling fluid is also characteristic of the effect of exothermic reaction involved in the setting of fresh cement behind casing [Conaway, 1987]. All of the spike anomalies mentioned in this paper occur in sections of borehole known to be uncased.

It should also be noted that the appearance of the anomalies shown in Fig. 1 is different if flow direction is reversed. For example, the form of anomaly type A is concave if water flows up the borehole and leaves at a fracture zone. An excellent example of this is shown by Drury et al. [1984, Fig. 4].

Each type of anomaly can be represented mathematically. Flow within a borehole was considered by Ramey [1962]; his simple model was based on the assumption that the vertical temperature gradient is negligibly disturbed by the flow. Ramey's model is expressed as:

$$v_z = v_0 + z\Gamma \pm [\exp(-z/A)-1]\ A\Gamma \qquad (1)$$

in which v_z is the temperature at depth z above or below the point of entry of the fluid to the borehole, v_0 is the temperature at the point of entry, Γ is the undisturbed thermal gradient, and A is a measure of the rate of heat transfer, and is a function of, among other parameters, thermal conductivity and thermal diffusivity of the rock penetrated by the borehole. Beck and Shen [1985] extended the analysis to allow for the vertical transfer of heat in the borehole. They found that the simple expression may lead to substantial errors in the estimation of temperature for very small (~hours) and very large (~decades) times after the onset of flow. Temperature is very sensitive to small flows. For example, the down-hole flow reported by Drury and Lewis [1983] in a Lac du Bonnet borehole was not detected by conventional hydrological techniques for several years; it was eventually traced to a crack in the surface casing.

Models of the storage of fluid in fractures were presented by Drury and Jessop [1982]. The phenomenon is modelled as the result of the liberation of heat from a continuous plane source, the basic equation of which is [Carslaw and Jaeger, 1959]:

$$v(x,t) = Q \int_0^t \exp[-(x-x')^2/4\alpha(t-t')]\ dt' / \sqrt{(t-t')} \qquad (2)$$

Equation 2 represents the distribution of temperature, v, with time, t, and distance, x, from a plane source (at x') in which heat is liberated at a constant rate $2\rho CQ\sqrt{(\alpha\pi)}$, where α is the thermal diffusivity of the rock, ρ is its density and C is its specific heat capacity. Drury and Jessop [1982] developed expressions for the source strength, Q, being constant during the period of heat exchange and linearly increasing during that period.

Lewis and Beck [1977] presented an expression for the effect of flow of fluid along a dipping fracture. The moving fluid carries heat, and therefore acts as a heat sink or heat source, depending on the direction of flow and the thermal gradient. By neglecting the effect of the finite dimensions of a fracture - rock body system, Lewis and Beck obtained a steady-state solution for the difference in apparent heat flow above (Q_u) and below (Q_l) the fracture:

$$|Q_u - Q_l| = fC\Gamma \sin\theta \qquad (3)$$

in which f is the mass flow rate of water, C is the specific heat capacity of water, Γ is the undisturbed thermal gradient, and θ is the angle of inclination (to the horizontal) of the fracture.

Expressions 1 to 3 require for their solution knowledge of the thermal conductivity and thermal diffusivity of the rock penetrated by the borehole from which the thermal anomalies are detected.

Display and Interpretation of Thermal Data

There are several ways in which the thermal perturbations can be displayed. In some cases a simple plot of temperature against depth is sufficient to show thermal anomalies. However, the sensitivity of thermal perturbations to small water flows [Drury et al., 1984] is such that enhancement of the display of temperature-depth data may be necessary to highlight thermal anomalies. For example, gradient changes can be enhanced by reducing the temperature-depth data by subtraction of a uniform gradient. An alternative method is to put the temperature data through a first difference filter to remove long wavelength trends. The data are then plotted as differential temperature against depth. For equispaced temperature measurements this is equivalent to displaying the gradient between successive points. This type of plot highlights small scale temperature variations that may not easily be recognized in temperature-depth plots that contain significant long wavelength information such as curvature due to climatic variations, or thermal perturbation remaining from the effects of drilling. Note that the analysis of changes of thermal gradient - for example, recognition of type C fractures as defined above - requires that the borehole be in thermal equilibrium, i.e. that drilling-induced disturbances to the gradient have dissipated.

Fig. 2 shows a temperature log of borehole URL-5 in three different

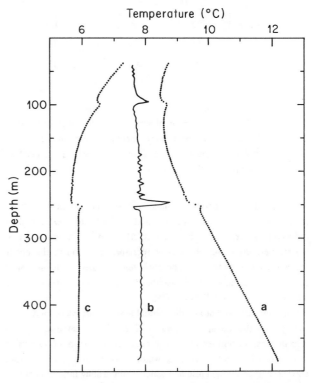

Fig. 2. Three representations of borehole temperature data: a) raw data; b) same data passed through first difference filter; c) data reduced by subtraction of gradient of 11 mKm^{-1}. Enhancements b and c highlight aspects of thermal anomalies, such as the offset below spike at 252 m, that are less apparent in the plot of raw data. Temperature scale is correct for raw data. Borehole is URL-5, Underground Research Laboratory, Manitoba.

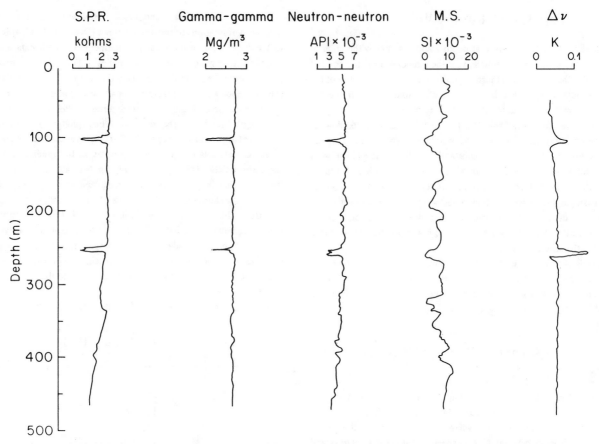

Fig. 3.   Comparison of some geophysical logs with differential temperature plot for borehole URL-5. Anomalies at 100 m and 252 m are clearly correlated among single point resistance (S.P.R.), density (Gamma-gamma), porosity (Neutron-neutron) and differential temperature (Δv) logs; correlation with smoothed magnetic susceptibility (M.S.) is positive, but less obvious.

ways: a) as true temperature against vertical depth, b) as differential temperature (Δv) against depth, and c) with temperature data reduced by subtraction of a gradient of 11 mKm^{-1}. The curvature in the temperature-depth plot and temperature inversion in the upper 150 m is a common feature of borehole temperature data from central and eastern Canada, and is ascribed to a climatic warming trend during the last century [Lewis, 1975].

The reduced gradient and differential temperature plots for URL-5 clearly show that the spike anomaly at 252 m is associated also with a step increase of temperature. The anomaly is interpreted as arising from a fracture that is receiving water flowing down the hole from above (i.e. type A), and also into which warm drilling fluid has entered (type B). There is no indication in the log of the depth at which water flows into the borehole; it is concluded that it must be close to the surface, and therefore undetected. There is also a gradient change at 252 m. The gradient is 9.7 ± 0.2 mK m^{-1} in the interval 200 - 240 m and 11.0 ± 0.1 mK m^{-1} from 260 m to bottom hole. There is no significant change in thermal conductivity of the rocks above and below the anomaly, with mean values of 3.60 ± 0.12 W m^{-1}K^{-1} and

3.55 ± 0.10 W m^{-1}K^{-1} respectively in the upper and lower intervals. The gradient change is interpreted as resulting from water flowing down a dipping fracture zone, removing heat such that conductive heat flow is lower, by 4 mW m^{-2}, above the fracture than below it. If the dip of the fracture is known independently, this heat flow difference can be used to estimate the flow rate of water along the fracture [e.g. Lewis and Beck, 1977].

The anomaly at 100 m shows only the characteristic effect of type B, i.e. an isolated fracture accepting drilling fluid, suggesting that it is not part of an extensive interconnected fracture system.

It is of interest to compare temperature logs with other geophysical logs. Fig. 3 shows the results of single-point resistivity, gamma-gamma (density), neutron-neutron (porosity), magnetic susceptibility and differential logs for borehole URL-5. The two major anomalies seen in the differential temperature log are well correlated with anomalies in electrical resistance, density and porosity logs that strongly suggest the presence of major fractures. The correlation with anomalies in the smoothed magnetic susceptibility log is significantly lower, but both fracture systems coincide with reduced magnetic susceptibility. This is

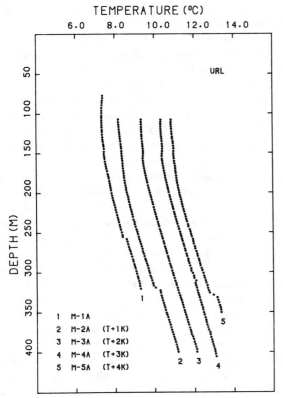

TEMPERATURE (°C)

Fig. 4. Temperature logs of M series boreholes at the Underground Research Laboratory.

characteristic of fractures in igneous rock, and it has been used as an indicator of sub-surface alteration [Chomyn et al., 1985].

The ease and speed with which temperature logs can be obtained highlights their usefulness in indicating the presence and type of fractures. As noted above, detailed analysis of thermal anomalies requires measurement of thermal conductivity, but simple temperature logs of holes in which drilling-induced thermal perturbations have dissipated, particularly in homogeneous bodies (e.g. plutons) in which major variations in conductivity are not expected, provide a powerful hydrogeological tool.

Thermal Logging at the URL Site

Introduction

The detection of fluid flow in crystalline rock is important for assessing the suitability of sites for the underground disposal of hazardous waste material. As part of the Canadian Nuclear Fuel Waste Management Programme (CNFWMP) temperature logging of several boreholes has been undertaken at the URL. Whenever possible, holes were logged several times at intervals of several weeks, in order that transient thermal anomalies, such as those associated with fracture type B and the disturbance caused by the drilling itself,

could be monitored. In addition, single logs have been run in some holes. It is noteworthy that most boreholes logged at the URL site showed some kind of thermal perturbation.

Thermal properties measurements have also been made on URL borehole core samples. The mean of 176 thermal conductivity values is 3.61 ± 0.40 Wm^{-1}K^{-1} (one standard deviation). The mean of 173 determinations of thermal diffusivity is 1.33 ± 0.24 mm^{2}s^{-1}. Combined with measurements of density, the thermal properties measurements yield estimates of specific heat capacity. The mean specific heat capacity of 173 samples is 1060 ± 206 Jkg^{-1}K^{-1}.

Figs. 2 and 4 illustrate some of the results obtained. Fig. 4 shows temperature logs of the M series large diameter holes. Prominent step increases in temperature are seen in the M-1A, M-2A and M-5A holes, whereas in the other two, possible thermal disturbances are much less easily discernible, particularly on the standard temperature-depth plot. The interpretation of each temperature log, based on the enhancement techniques outlined above, is summarized in Table 1. The plots of Figs. 2 and 4 serve to indicate the ambiguities and uncertainties that arise in such interpretations. For example, the thermal perturbation in M-3A at 380 m, suggesting a type A anomaly is not seen clearly in the usual temperature-depth plot. In the following discussion, all analyses have been undertaken using the three different means of displaying data identified above.

Mapping Fracture Systems

The temperature logging was undertaken as part of a major geophysical/hydrological study and so was subjected to several restrictions with regard to accessibility of boreholes, particularly at times when they were free of disturbances induced by other logging or pumping activities. It was not possible, therefore, to obtain data on a regular basis, nor to repeat some logs that gave ambiguous data because they were obtained - of necessity - too soon after previous down-hole experiments. It is beyond the scope of this paper to present all the data in detail. The interpretations are summarized in Table 1, in which those that are ambiguous are clearly indicated.

TABLE 1. Major fractures, characterized by type (see text), detected in URL boreholes by temperature logging. Question marks indicate uncertain interpretations.

Hole	Depth (m) and type of fracture		
URL-1	111 (B)	180 (A?)	312 (A?)
URL-2	163 (A)		
URL-3	70 (B)	150 (A?)	
URL-5	100 (B)	252 (A+B+C)	
URL-6	68 (B?)	125 (B)	272 (A+B)
URL-7	71 (B)	143 (B?)	
M-1A	145 (B?)	259 (A)	
M-2A	323 (A)		
M-3A	155 (B)	380 (A)	

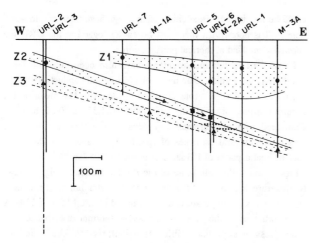

Fig. 5. Interpretation of Underground Research Laboratory site borehole temperature logs in terms of fracture zone distribution. Arrows indicate postulated direction of water flow. Symbols for fractures are: solid circle - fracture receiving drilling fluid but otherwise exhibiting no thermal indications of fluid flow; triangle - fracture into which water is flowing from the borehole; square - fracture into which water is flowing from the borehole and at which there is also a change in gradient not associated with a change in thermal conductivity. Parallel dotted lines indicate a possible connection between zones Z2 and Z3 postulated from hydrological data.

The number and distribution of boreholes logged at the URL site permits fracture mapping to be attempted. Fig. 5 shows a west-east profile onto which have been projected the true vertical depths of those boreholes that are within 100 m of the profile at the surface. Fracture types interpreted from the temperature logs are indicated symbolically. In the case of fractures exhibiting more than one of the possible thermal anomalies, the symbol in Fig. 5 for that fracture shows the larger or largest scale flow type. The three zones shown in Fig. 5 are based on integrated interpretations from surface geophysical work, from borehole hydrological tests, and from the interpretations of the temperature data.

Z1 is a zone between approximately 50 m and 150 m depth. The extent of the zone, both laterally and vertically is poorly defined [Davison et al., 1982]. Most fractures in the zone indicated from thermal logs are of type B, suggesting that they are generally unconnected. As the thermal anomaly arising from entry of drilling fluid into a fracture is governed in part by the ease with which the fluid can enter, the magnitude of the anomaly is related to the hydraulic conductivity of the fracture in the vicinity of the borehole. In URL-5 the spike thermal anomaly at 252 m can be modelled by a uniform heat source of strength 2 Kms^{-1} acting for a period of 10 days, for a total heat input to the fracture of 3 MJm^{-2}, or an average heat input rate of approximately 3.5 Wm^{-2}. A similar calculation for the URL-5 spike anomaly at 100 m gives 6 Wm^{-2}, which suggests that the deeper

fracture has a lower hydraulic conductivity at the point at which it is penetrated by the borehole.

Z2 is characterized by few thermal indications of fractures. The zone dips eastward at approximately 18° [Green and Soonawala, 1982]. There are indications from both URL-5 and URL-6 of flow down the dipping zone. The changes with depth of temperature gradients in the boreholes URL-5 and URL-6 imply down-dip water flow. As noted above, the difference in heat flow above and below the fracture at 252 m in URL-5 is 4 mWm^{-2}. Taking the thermal gradient below the fracture zone as the undisturbed value, the rate of flow of fluid in the zone is estimated to be 0.3 $gs^{-1}m^{-1}$.

Z3 is characterized by thermal anomalies indicative of fractures that are receiving water flowing down the boreholes. It also dips eastward, at approximately 14°. Davison et al. [1982] postulated a connection between Z2 and Z3 on the basis of hydrological data , approximately in the position indicated in Fig. 6. The rate of flow down borehole URL-5 is estimated to be approximately 2.5 x 10^{-5} m^3s^{-1}.

It is noteworthy that each zone defined independently of the thermal data contains generally uniform fracture types as indicated by their thermal signatures. For example, Zone 1 is characterized by fractures that do not seem to be well connected. The inference from this is that the thermal data provide information on the hydrological uniformity of the zones, and the potential coupling between them. Type A anomalies indicate flow between zones that are at different hydraulic potentials. Such zones might not normally have any connection, a factor that must be considered in the construction of any kind of underground facility. It is concluded that thermal logging, in conjunction with thermal properties measurements, offer a valuable contribution to a multidisciplinary study of crystalline rock bodies. Not only do temperature logs indicate the presence of fractures, but they provide some indication of the direction and rate of water flow.

Acknowledgments. I thank Drs. A.E. Beck and A.M. Jessop for reviews of this paper. Contribution of the Geological Survey of Canada no. 47887.

References

Beck, A. E., and P. Y. Shen, Temperature distribution in flowing liquid wells, *Geophysics, 50*, 1113-1118, 1985.

Carslaw, H. C., and J. C. Jaeger, *Conduction of heat in solids*, Clarendon Press, Oxford, 2nd ed., 510pp., 1959.

Chomyn, B. A., W. A. Morris, P. Lapointe, and R. L. Coles, Application of magnetic susceptibility to assessing the degree of alteration of crystalline rock, *Atomic Energy of Canada Ltd. Tech. Rec. TR-299*, 609-621, 1985.

Conaway, J. G., Temperature logging as an aid to understanding groundwater flow in boreholes, *Los Alamos National Laboratory Paper LA-UR-87-3355*, 9 pp., 1987.

Davison, C. C., A. Brown, and N. M. Soonawala, Preconstruction site evaluation program at the Canadian Underground Research Laboratory, *Atomic Energy of Canada Ltd. Tech. Rec. TR-207*, 162-187, 1982.

Drury, M. J., and A. M. Jessop, The effect of a fluid-filled fracture on the temperature profile in a borehole, *Geothermics*, 11, 145-152, 1982.

Drury, M. J., and T. J. Lewis, Water movement within Lac du Bonnet batholith as revealed by detailed thermal studies of three closely-spaced boreholes, *Tectonophysics*, 95, 337-351, 1983.

Drury, M. J., A. M. Jessop, and T. J. Lewis, The detection of groundwater flow by precise temperature measurements in boreholes. *Geothermics*, 13, 163-174, 1984.

Fyfe, W. S., Fluids in deep continental crust, in *Reflection seismology: the continental crust, Geodyn. Ser.*, vol. 14, edited by M. Barazangi and L. Brown, AGU, Washington, D.C., 33-39, 1986.

Gale, I.N., and R. A. Downing, Heat flow and regional groundwater flow in the United Kingdom, *Brit. Geol. Survey. Rep.*, 22 pp., 1986.

Gatenby, R.M., Conduction of heat from sheep to ground. *Agricultural Meteorology*, 18, 387-400, 1977.

Green, A. G. and N. M. Soonawala, Detection of fracture zones in granites by geophysical logging and seismic reflexion surveys, *Proc. Workshop on Geophysical Investigations on Connection with Geological Disposal of Radioactive Waste*, Nuclear Energy Agency, Ottawa, 21-36, 1982.

Hoisch, T. D., Heat transport by fluids during Late Cretaceous regional metamorphism in the Big Maria Mountains, southeastern California, *Geol. Soc. Amer. Bull.*, 98, 549-553, 1987.

Kerrich, R., Fluid transport in lineaments, *Phil. Trans. R. Soc. Lond.*, A317, 219-251, 1986.

Kozlovsky, Y. A., The world's deepest well, *Scientific American*, December, 98-104, 1984.

Lewis, T. J., A geothermal survey at Lake Dufault, Quebec, *Ph.D. thesis*, University of Western Ontario, London, Ontario, 1975.

Lewis, T. J. and A. E. Beck, Analysis of heat flow data: detailed observations in many holes in a small area, *Tectonophysics*, 41, 41-59, 1977.

Majorowicz, J. A. and A. M. Jessop, Regional heat flow patterns in the Western Canada Sedimentary Basin, *Tectonophysics*, 74, 209-238, 1981.

Majorowicz, J. A., F. W. Jones, and A. M. Jessop, Geothermics of the Williston Basin in Canada in relation to hydrodynamics and hydrocarbon resources, *Geophysics*, 51, 767-779, 1986.

Mansure, A. J. and Reiter, M., A vertical groundwater movement correction for heat flow, *J. Geophys. Res.*, 84, 3490-3496, 1979.

Ramey, H. J., Well bore heat transmission, *J. Petrol. Technol.*, 14, 427-435, 1962.

Rasmusen, A. and Neretnieks, I., Radionuclide transport in fast channels in crystalline rock, *Water Resources Research*, 22, 1247-1256, 1986.

Stegena, L., Water migration influences on the geothermics of basins, *Tectonophysics*, 83, 91-99, 1982.

TERRESTRIAL HEAT FLOW VARIATIONS IN THE NORTHEASTERN PART OF THE STATE OF SAO PAULO: A CASE FOR TRANSPORT OF GEOTHERMAL HEAT BY INTERFRACTURE FLUID FLOWS

A. C. del Rey

Instituto Astronomico e Geofisico (USP), Caixa Postal 30.627, Sao Paulo, Brazil

V. M. Hamza

Instituto de Pesquisas Tecnologicas (IPT), Caixa Postal 7141, Sao Paulo, Brazil

Abstract. Terrestrial heat flow density (HFD) measurements have been made at sixteen localities, in the Precambrian metamorphic terrain belonging to the Ribeira fold belt, in the northeastern part of Sao Paulo. Results obtained show that the HFD varies substantially, in the range 33 mW m^{-2} to 103 mW m^{-2}, over distances of a few tens of kilometers. Since available geophysical data do not indicate the presence of intrusive bodies or structural discontinuities capable of generating large scale variation in HFD within this area, the possibility that such changes are induced by groundwater flows through local fracture systems was considered. In order to examine in detail the influence of fracture systems on the local geothermal regime, a small area of 26 km x 42 km was selected for detailed study within the region. Aerial photographs were used in the identification of more than 2000 fractures within the study area. Analysis of fracture parameters revealed that the orientation and degree of interconnection of fracture systems exert strong influences on the hydrological characteristics of water wells and spatial distribution of thermo–mineral springs. To examine the nature of the influence of interfracture fluid flows in modifying the local geothermal regime, "permeable" fractures were selected on the basis of data on pumping tests of water wells. Thermally anomalous zones in the vicinity of such fracture planes were delimited on the assumption that lateral heat flow would be significant to distances of the order of the length of the fractures. Spatial distribution of such zones indicates that over fifty per cent of the area is affected by interfracture fluid flow. Examination of heat flow data within the study area shows that "cold zones" in the vicinity of fracture planes are characterized by low HFD values. A systematic tendency of increasing HFD with distance from the fracture zones is also observed. The results indicate that conventional measurements uncorrected for interfracture fluid flow effects may be underestimating HFD in continental hard rock terrains by as much as 10 to 20 percent.

Introduction

It is well known that near surface layers of the earth contain a host of discontinuities such as faults, fractures, fissures and joints, whose dimensions may vary from a fraction of micron to several hundred kilometers. Discontinuities such as faults and fractures arise from the action of mechanical forces associated with tectonic activities while small scale discontinuities are the result of weathering or alteration processes acting on the inhomogeneous rock mass. Fluids entering such discontinuities can penetrate to depths depending on their permeability variations and the degree of interconnection. The flow paths are generally complicated and the magnitudes of such flows decrease rapidly with depth. In many cases there are indications that flows exist to depths of several kilometers in the upper crust. Associated with such flows is the advective heat transport, and there is no doubt that deep flows are a very efficient means of scavenging geothermal heat from depths. The ascending fluids discharge themselves into shallow aquifers or natural drainage systems. In cases where this discharge leads to significant temperature changes the presence of circulation systems becomes noticeable easily. However, if the temperature differences are small, the presence of circulation systems may not be noticed even though substantial quantities of heat are being discharged. It is quite possible that in terrains of high fracture density, discharge of geothermal heat by fluid flows at low temperature differences is ubiquitous. Detection of such flows is nevertheless likely to be difficult because of the dominating influence of other factors that control the heat budget of near surface layers and drainage systems.

Copyright 1989 by
International Union of Geodesy and Geophysics
and American Geophysical Union.

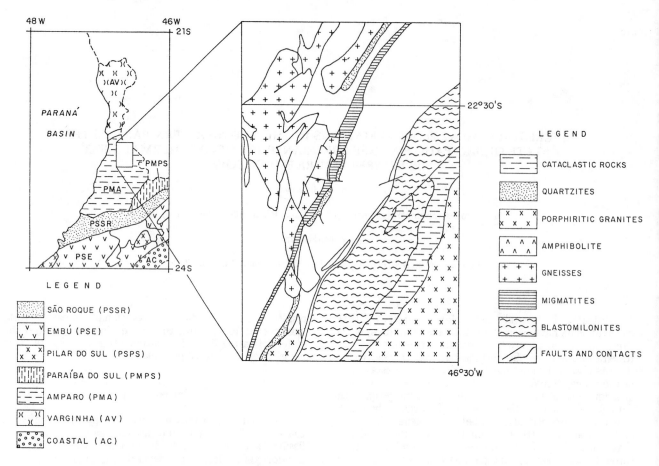

Fig. 1. Simplified geologic maps of northeastern Sao Paulo (left) and the study area within it (right). The legend on the left refers to the main stratigraphic units in northeastern Sao Paulo.

In the present context, the interest is in evaluating the relative importance of fracture controlled advective heat transfer on a regional scale. It is fairly well established that low conductive HFD values encountered in young oceanic crust are a result of heat loss to sea water circulating through fractures in basaltic lava flows which constitute the oceanic basement. In old ocean crust under the cover of thick sedimentary layers, sea water does not have easy access to the fractures and conductive HFD is believed to represent the total heat flux. For continental regions, however, the emerging opinion among geothermal researchers is just the opposite. Groundwater flow is believed to transport a small but significant quantity of geothermal heat in continental sedimentary basins, while Precambrian hard–rock terrains are assumed to be free of such perturbing effects. The main objective of the present paper is to examine the limits of this assumption in the light of geothermal investigations within an area where conditions exist for evaluating the nature of discontinuities on a convenient scale. Specifically we consider the nature of fluid flows through fracture systems within a Precambrian metamorphic terrain in the northeastern part of the state

of Sao Paulo and its influence on the local geothermal regime.

Geology of the Study Area

The Precambrian metamorphic terrain in the north–eastern part of the state of Sao Paulo lies within the so–called Ribeira fold belt. The last major thermo–tectonic event that affected the region is considered to be the Brasiliano event during the period 450–700 Ma. This fold belt is limited by the Sao Francisco craton to the north and Phanerozoic sediments of the Parana basin to the west. The area selected for the present study is approximately a rectangular strip between latitudes 22°25'S and 22°45'S and longitudes 46°30'W and 46°46'W. Simplified geologic maps of northeastern Sao Paulo and the study area within it are shown in figure (1).

The main stratigraphic units occurring within the study area are the Amparo and Varginha complexes with ages of Proterozoic and Archean respectively (Hasui et al, 1981). The Varginha complex is limited to the northeastern corner of the study area. The main rock types constituting the Amparo complex are granites, gneisses, migmatites, granitoids and quartzites. The migmatites occur in the central region separating gneisses and granites in the north and northwest and granitoids in the southeast. There are several elongated quartzitic bodies in the central parts extending in the NE–SW direction. Also large strips of cataclastic rocks are found associated with the faults.

The most conspicuous structural features in the region are the immense transcurrent and thrust faults extending across the entire northeastern part of the state of Sao Paulo, for distances of several hundred kilometers. A good number of them have the general orientation NE–SW and cut across the fold belts and post–tectonic granitoids of the upper Proterozoic. Most of the faults have been reactivated during Mesozoic and Cenozoic time (Hasui et al, 1980). The structural evolution of the region is complex and several attempts have been made to elucidate the sequence of tectonic events (Almeida et al, 1973); Hasui et al, 1975; Fiori et al, 1978; Almeida et al, 1980). Campanha et al (1983) identify five different phases of folding in which the last two events, identified as Ft+3 and Ft+4, have produced extensive fracture systems with directions NNE–SSW and N–S, and ESE–WNW and E–W respectively. Within the study area the Ft+3 event with fractures in the direction NNE–SSW is dominant. The main faults cutting across the study area are the Monte Siao and Ibitinga both with NNE–SSW direction and Jacutinga with WNW–ESE direction.

Identification of Fractures and Fracture Systems

Aerial photos taken over the study area were used in the present work for identification of fractures. The photographs were examined under a zoom stereoscope for marking linear features. Standard procedures employed in photogeologic studies (Avery, 1962; Ricci and Petri, 1965) were adopted for fracture identification. Linear portions of streams, rivers and drainage systems as well as specific topographic features and changes in vegetation cover are generally indicative of fractures. However, access to topographic maps and some previous knowledge of the geology and morphology of the area under consideration are of great help in the final identification of fractures. It should be noted that only those fractures that intersect the surface, or have a surface expression, can in principle be identified in air–photos. Horizontal fractures cannot be seen and sub–horizontal fractures can hardly be identified in air–photos because of their subdued surface expression.

More than 2000 fractures were identified within the study area. Only those more than 500 meters long were considered in the present work. All fractures identified in air–photo studies were marked in topometric maps on the scale 1:50,000 for further analysis. Many of the fracture systems identified in air–photos were checked against observations made during recent geologic mapping of the area at the scale 1:50,000 (Campanha et al, 1983; Grossi Saad and Barbosa, 1983). Some of the identifications in air–photos were checked against direct observations in the field, at sites near water wells which

were visited for geothermal studies. The main fracture parameters that can be discerned in air–photos are the location, orientation and length. Figure (2) provides a general view of the distribution of fractures thus identified. For studying the frequency distribution of fracture orientations, original fracture maps at the scale 1:50,000 were subdivided into 4 km x 4 km squares and fractures divided into groups based on their orientations, with class intervals of 10°. A template was used for this purpose. The results obtained are presented in the form of rose diagrams of fracture directions in figure (3). The principal orientations of fracture systems in decreasing order of importance are N70–79E, N60–69W, N10–19E and N40–49W.

Relations between Fracture Systems and Subsurface Flows

We are interested in studying the influence of fracture systems on fluid flows that are capable of affecting the local geothermal regime. For this purpose it was found

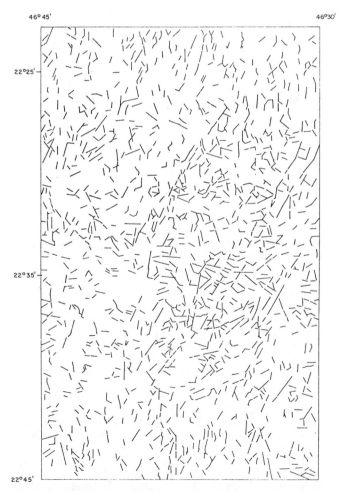

Fig. 2. Distribution of fractures identified in air–photo studies within the study area.

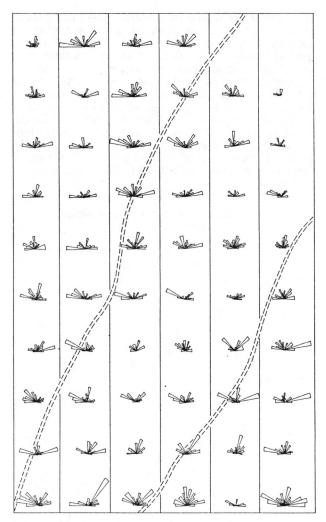

Fig. 3. Rose diagrams of fracture directions determined for 28 localities within the study area. The broken lines indicate faults.

measured. The wells were divided into groups depending on the orientation of the local fracture systems, with class intervals of 20°, and their data on flow rates and specific capacities compared. The results obtained are shown in table (1). High values of specific capacities, which are indicators of good permeability, were found to be associated with fractures having orientations N0–20E, N21–40E and N61–80E. Though the data set is insufficient to define the nature of flows and determine hydrological parameters, there is no doubt that fractures having these orientations are characterized by relatively high permeabilities. It is reasonable to expect that groundwater flows through these fractures are contributing to removal of geothermal heat from depths.

Lack of wells penetrating to depths greater than 200m prevents us from making a similar study of the deep flow systems. However, some information on the nature of fluid circulation at depth can be inferred by evaluating the degree of interconnection of fractures and its association with the spatial distribution of thermo—mineral springs. Studies carried out by Snow (1965), Sagar and Runchal (1977), Long (1983) and Witherspoon (1986) show that the degree of interconnection in fracture systems is a function of the fracture density which in turn is related to the overall hydraulic permeability of the medium. In order to evaluate the density of fractures the fracture map was subdivided into 2 km x 2 km squares and the total length of fractures calculated within each square. Total fracture length divided by the area of the square was called the length density of fractures (LDF). Values of LDF were plotted in the centre of each square for drawing iso—LDF curves. To investigate the influence

convenient to disregard flows taking place in the first few tens of meters as they are directly related to the surface drainage systems and are of secondary importance in the present context. The remaining flow regime can be considered as falling into two parts: 1) a shallow system whose characteristics can be determined from data on pumping tests of water wells in the area and 2) a deep system, some features of which can be inferred from the spatial distribution and chemical characteristics of thermo—mineral springs. By shallow flow system we mean flows taking place at depths ranging from a few tens to a few hundreds of meters. In a deep system flows may penetrate to depths of several kilometers.

Data on pumping tests of 102 water wells were collected and those which provided reliable data on both flow rate and specific capacity were selected for further analysis. The sites of such wells were visited subsequently and orientation of local fracture systems

TABLE 1. Data on flow rates and specific capacities of water wells grouped into classes based on orientations of associated fractures. (Nf is number of flow rate measurements, F mean flow rate, σf standard deviation of F, Nc the number of specific capacity measurements, C the mean capacity and σc the standard deviation of C).

Dip	Strike	Flow Rate (m³ h⁻¹)		Specific Capacity (m² h⁻¹)	
		Nf	F±σf	Nc	C±σc
NW	NS–N20E	4	5.8±3.7	2	0.16±0.14
	N21–N40E	2	6.2±2.9	2	0.17±0.14
	N41–N60E	8	2.0±1.1	2	0.06±0.01
	N61–N80E	9	2.7±1.7	6	0.05±0.04
	NS–N20W	3	2.6±1.0	2	0.09±0.10
	N21–N40W	4	3.3±3.4	4	0.08±0.06
	N41–N60W	2	0.7±0.2	1	(0.01)
	N61–N80W	3	2.1±2.2	3	0.04±0.04
SE	NS–N20E	5	1.7±1.4	1	(0.04)
	N21–N40E	–	–	–	–
	N41–N60E	5	2.3±1.7	1	(0.03)
	N61–N80E	5	2.4±0.8	3	0.16±0.20
	NS–N20W	2	1.5±0.2	2	0.02±0.004
	N21–N40W	1	(1.1)	1	(0.04)
	N41–N60W	8	1.9±1.6	3	0.09±0.04
	N61–N80W	1	(1.7)	1	(0.02)

△   thermal springs

—24— fracture intensity (Km/Km^2)

Fig. 4. Map of length density of fractures (LDF). The interval used for contouring is 4 km km^{-2}.

of LDF on deep flows, data on 52 thermo–mineral springs occuring at seven localities were collected and plotted on the map of LDF. The results, presented in figure (4), show that there is a reasonably good correlation between LDF and the occurrence of thermo–mineral springs. Most of the springs are located in areas characterized by LDF values higher than 16 km^{-1}. Also all springs with flow rates in excess of 60 litres per minute are localized in regions with LDF values higher than 20 km^{-1}. The results thus indicate that fracture densities calculated from air–photo studies are reasonably good indicators of flows at depth.

Geothermal Investigations

HFD measurements were made at twenty sites in the northeastern parts of the state of Sao Paulo of which seven, distributed amongst four localities, are within the study area. Of the remaining sites, nine are located in the surrounding Precambrian regions, and four in the Paranha basin to the west of the study area.

Results of in–hole temperature logs in thermally stabilized wells, in combination with thermal conductivity measurements on representative rock samples, were used in calculating HFD values. Calibrated thermistor probes attached to light–weight three–conductor cables were used for temperature measurements in boreholes and wells. The accuracy of the logging system is better than 0.1°C while relative differences can be measured with a precision of 0.01°C. Temperatures were measured while the probe was being lowered into the well. The interval chosen for measurements was usually two meters. Corrections for drilling disturbances were considered unnecessary as the time elapsed between drilling and logging operations were of the order of several months. Temperature logs revealed marked curvatures in the upper 50–100 meters which are usually characteristic of the effects of recent changes in the mean annual surface temperatures. Attempts to model the climatic history indicated that the results are compatible with increases in the mean annual surface temperatures of the order of a few degrees in the last 50 to 100 years. Such changes may be indicative of a general climatic warm–up triggered by natural or man–made activities. In any case the effects of such changes on geothermal gradients at depths of more than 100 meters were found to be less than 5%. Also perturbing effects of local topography on measured temperatures were calculated, but the magnitude of corrections was found to be less than 5% at depths exceeding 100 meters. Corrections for water flows within wells were not considered because temperature–depth plots were fairly linear and free of perturbations characteristic of such flows. As typical examples, vertical distributions of measured and corrected temperatures for the three sites in the locality of Serra Negra are presented in figure (5).

Thermal conductivity measurements were carried out using divided–bar and half–space line source apparatus, calibrated using standard discs of known thermal conductivity. For divided–bar measurements samples were cut into the form of discs with dimensions identical to those of the standard discs, while for the half–space line source method, preparation of a single flat and polished surface of adequate dimensions was found to be sufficient, provided the sample is isotropic and homogeneous. In the case of drill cuttings disc shaped sample holders were used and measurements carried out under water saturated conditions. Thermal conductivity of the solid fraction was then calculated from that of the water saturated mixture using models of distribution of constituent phases. Geometric (Woodside and Messmer, 1961), Bruggeman (Hutt and Berg, 1968) and Hashin–Shtrikmann (Horai and Simmons, 1969) models were employed and the procedure suggested by Marangoni (1986) used for eliminating inconsistent results. For some wells neither core samples nor drill cuttings were available. In such cases conductivity measurements were carried out on fresh samples collected from outcrops of rock formations that are representative of the main lithologic units at depth. In the case of

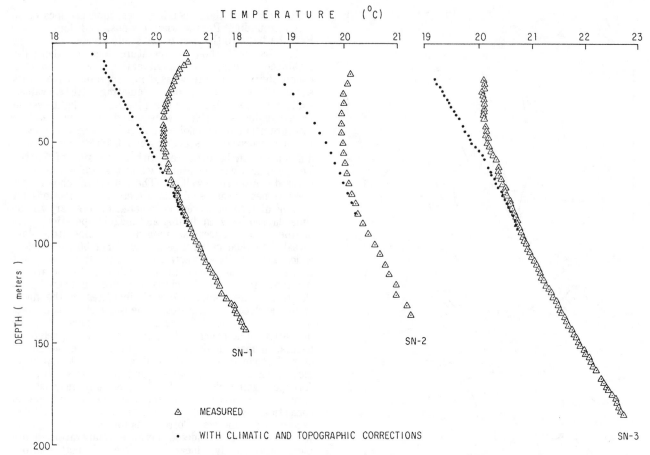

Fig. 5. Vertical distribution of measured and corrected temperatures for three wells in the locality of Serra Negra.

anisotropic rocks (migmatites, gneisses, quartzites and mylonites) conductivity was measured in directions parallel and perpendicular to the bedding planes or foliations. Details of such measurements are reported in Marangoni and del Rey (1986). A total of 234 measurements were made. Mean thermal conductivities of the main rock types encountered within the study area are given in table (2).

A summary of the gradient, conductivity and HFD values for the four localities (7 borehole sites) within the study area is given in table (3). A typical result showing the temperature profile, interval gradients and thermal conductivities and Bullard plot for the locality of Monte Alegre do Sul, within the study area, is presented in figure (6). Table (3) gives the results of HFD measurements in eight localities in the surrounding Precambrian regions and four in the Parana basin to the west of the study area. These data provide a general idea of the regional HFD pattern. With the exception of Jaguariuna and Cosmopolis the HFD in neighbouring regions outside the study area is quite uniform. The geothermal gradient value for Jaguariuna is doubtful because of strong indications of water flow within the well. The HFD value for Cosmopolis, located within the

Parana basin, seems to be reliable but is an isolated value and the geological context is not representative of the conditions in the northeastern parts of the state of Sao Paulo. The surprising fact, however, is that within the

TABLE 2. Thermal conductivities of main rock types encountered within the study area. (N is the number of thermal conductivity measurements, λ the mean value in W $m^{-1}K^{-1}$, A the anisotropy ratio and σ the standard deviation).

Rock Type	N	$\lambda \pm \sigma$	$A \pm \sigma$
Migmatite	45	3.4±0.7	1.2±0.3
Blastomilonite	5	3.5±0.3	1.2±0.1
Granitoide	34	3.0±0.6	—
Porphyritic Granite	5	3.5±0.3	—
Granitic Gneisse	23	3.2±0.4	1.2±0.2
Quartzite	97	5.0±1.0	1.0±0.2
Sandstone	29	3.9±1.7	—
Siltstone	6	3.0±0.5	—
Argilite	12	2.1±0.5	—

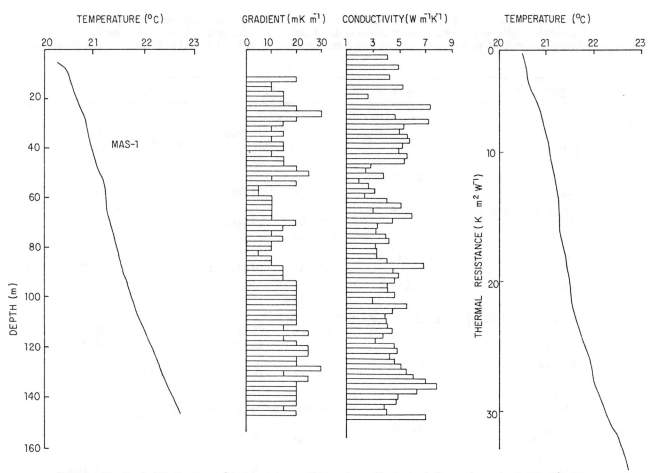

Fig. 6. Vertical distribution of temperatures, interval gradients and thermal conductivities for the well MAS–1 in the locality of Monte Alegre do Sul.

study area, where the geothermal regime is expected to be uniform, the HFD varies substantially, from a low of 33 mW m⁻² to a high of 103 mW m⁻², within distances of a few tens of kilometers. Other available geophysical information (aeromagnetic and gravity data) does not indicate the existence of intrusive or other geological discontinuities capable of producing such large–scale variations in heat flow within the study area. The possibility that such variations are induced by ground water flow through extensive networks of fracture systems will now be considered.

Heat Transport by Interfracture Fluid Flows

Meteoric water flowing through fractures would cool the surfaces of fracture planes and this cooling effect can penetrate to considerable distances within the rock mass. The magnitude and penetration of the cooling effect would depend on the flow rate, geometry of the flow path and thermal properties of the rock mass. For interconnecting fracture systems the geometry of the flow path is highly complicated, and simplifying assumptions

are necessary for modeling thermal effects of flows. Simplified models of heat transfer in pipe flow (Turcotte and Schubert, 1982) show that if the flow is weak the geothermal regime is hardly perturbed whereas under moderate flows the descending limb of the circulation path is cooled while the upper part of the ascending limb is heated. In cases where the flow is substantial the entire flow circuit will be affected by cooling. It is natural to expect strong horizontal gradients and lateral flow of heat in the vicinity of fracture planes. Close to such fracture systems vertical heat flow will be altered substantially and true HFD can be measured only at considerable distances from the fracture planes. In other words one may expect systematic variations of HFD with distance from the fracture zone and the existence of such a trend is a definite indication of the presence of heat transport by interfracture fluid flows.

In order to evaluate the magnitude of cooling effects within the study area it was necessary first to identify fracture systems in which flows are taking place and then to estimate, by some means, the distance to which cooling effects have penetrated. Since there was no easy way of directly identifying fractures which are

TABLE 3. Summary of geothermal gradient, thermal conductivity and heat flow density values for 20 localities in the northeastern parts of the state of Sao Paulo (ΔZ is depth interval in meters, Γ is geothermal gradient in mK m^{-1}, λ is thermal conducivity in W m^{-1} K^{-1} and Q is heat flow density in mW m^{-2}).

Localities/Site	Coordinates	ΔZ	Γ	λ	Q
Serra Negra	22°37'/46°42'				
SN—01		136—184	23.82±0.02	3.6±0.7	86±17
SN—02		105—143	22.71±0.05	3.0±1.0	68±23
SN—03		85—135	20.30±0.10	3.0±1.0	61±20
Monte Alegre Sul (1)	22°43'/46°43'				
MAS—01		89—147	20.57±0.01	5.0±1.0	103±21
Aguas de Lindoia (1)	22°29'/46°39'				
AL—01		135—199	19.44±0.01	2.9±0.4	56±8
AL—02		124—142	17.30±0.40	2.9±0.4	50±7
Lindoia	22°31'/46°39'				
LD—01		44—94	10.93±0.02	3.0±1.0	33±11
Itapira (2)	22°26'/46°49'				
IP—01		97—143	15.58±0.03	3.0±1.0	47±16
Braganca Paulista(2)	22°57'/46°33'				
BP—01		150—178	21.74±0.04	3.0±1.0	65±22
BP—02		94—178	22.78±0.03	3.0±1.0	68±27
Atibaia (2)	23°07'/46°33'				
AT—01		60—150	13.74±0.01	3.1±0.6	43±8
Pedreira (2)	22°45'/46°54'				
PD—01		40—90	10.09±0.03	3.0±1.0	30±10
Amparo (2)	22°42'/46°47'				
AM—01		42—204	18.06±0.01	3.0±1.0	54±18
Jundiai (2)	23°12'/46°52'				
LD—01		80—172	19.11±0.08	3.0±1.0	57±19
Itu (2)	23°16'/47°18'				
IT—02		61—177	17.25±0.04	3.0±0.8	67±14
Jaguariuna (2)	22°42'/46°59'				
JG—01		130—150	(34±14)	(3±1)	(102±54)
Mogi—Mirim (3)	22°26'/46°57'				
MN—01		42—120	14.89±0.06	4.0±2.0	60±30
Araras (3)	22°22'/47°23'				
AR—01		61—117	13.30±0.40	2.7±0.6	36±8
Rafard (3)	23°00'/47°31'				
RF—01		72—120	18.95±0.07	2.1±0.1	40±2
Cosmopolis (3)	22°38'/47°12'				
CO—01		120—292	30.23±0.01	3.9±0.8	118±24

"permeable", the simplifying assumption was made that fractures associated with high productivity groundwater wells be considered as "hydrologically active". Following this assumption, and the information shown in table (1), fractures with orientations N0–20E, N21–40E and N61–80E were selected. The remaining fracture systems were considered as "hydrologically inactive" and eliminated from further analysis. Next, to eliminate the complications involved in calculating the penetration distance, it was assumed that the cooling effect would be significant to distances of the order of the length of the fracture; thus circles drawn around each fracture with fracture length as diameter would provide an estimate of the area affected by cooling. Figure (7) shows the results obtained by drawing circles around "permeable" fractures within the study area. As can be seen from this figure, in

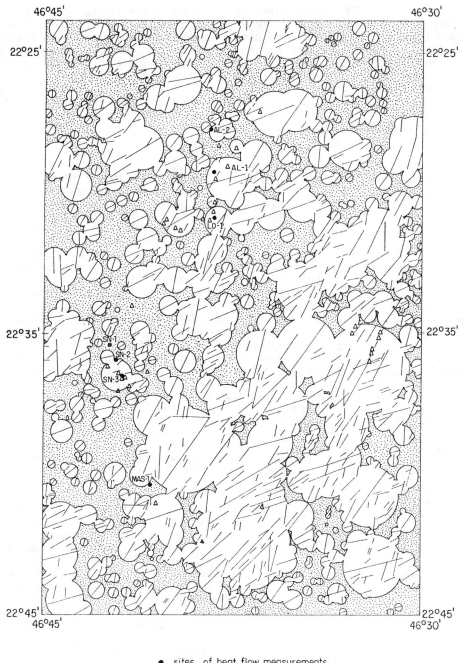

Fig. 7. Distribution of "permeable" fractures and associated "cold zones" within the study area.

areas of high fracture density intersecting circles would lead to extensive "cold zones" affected by lateral heat transport.

The distribution of HFD values within the study area was now examined to verify if there are systematic variations in relation to zones of lateral heat flow. Of the four localities two (Aguas de Lindoia and Lindoia) are within zones affected by cooling while one (Monte Alegre

TABLE 4. Comparison of mean HFD values within the study area and neighbouring regions in the northeastern parts of the state of Sao Paulo. The doubtful HFD value for Jaguariuna has been omitted in calculating the mean for Precambrian regions outside the study area. (N is the number of HFD measurements, Q the mean HFD in mW m^{-2} and σ the standard deviation of Q.).

Description of the Area	N	Q$\pm\sigma$
Study Area	7	65±23
– "Cold Zones"	5	54±13
– "Warm Zones"	2	94±12
Neighbouring Regions	13	61±25
– Precambrian Terrain	8	54±13
– Parana Basin	4	64±38
Northeastern Sao Paulo	20	62±24

do Sul) is outside. In Serra Negra sites of wells SN–2 and SN–3 are within the "cold zones" while SN–1 is free of cooling effects. An examination of HFD values (see Table 3) reveals that sites within "cold zones" are characterized by relatively low HFD values in relation to those outside it. Comparison of mean HFD values for the "cold" and "warm" zones within the study area and the neighbouring regions, given in table (4), provides an indication of the magnitude of lateral heat flow in the vicinity of permeable fracture planes. A plot of HFD versus distance from the centre of the nearest fracture zone, presented in figure (8), shows that the results are compatible with postulated effects of interfracture fluid flows.

The high HFD at the two sites, MAS–1 and SN–1, may be considered anomalous in relation to the regional HFD pattern. It is possible that these two sites are situated close to fracture systems not identified in air–photo studies, and which permit upward flow of

warm water from deeper levels. In such cases, associated heating effects can produce higher than normal heat flow in the surrounding rock mass. The only other possible mechanism capable of producing local HFD anomalies is the thermal refraction effects at the contact zone of two lithologic units of differing thermal conductivity (Jones and Oxburgh, 1979). Though such a possibility cannot entirely be ruled out the mechanism involving upflow of warm waters seems to be more likely in the present context.

Discussion and Conclusions

According to current practice in geothermal research, corrections for perturbing effects of water flows are made only in cases where thermal effects of flow are clearly discernible in temperature logs. The results obtained in the present work show that it may be necessary to consider perturbing effects of flow through nearby fracture zones, even in cases where there are no indications of flow either in the well or in the rockmass surrounding it. Thus isolated HFD measurements in areas of high fracture density may not be representative of true HFD.

With the exception of the work by Lewis and Beck (1977), there have been few studies with the objective of examining the nature of HFD variations on a local scale. The present work is an attempt to examine local variation of HFD in terms of heat transport by water flows through fracture zones. Techniques of fracture analysis were employed for this purpose and the results obtained show that local variations of HFD within the present study area are compatible with postulated effects of groundwater flow through networks of interconnecting fracture systems. It is perhaps significant to note that similar conclusions were reached by Lewis and Beck (1977) in their study of local variations of HFD in a small area in the Canadian Shield.

Since cooling effects can penetrate considerable distances into the impermeable rock formations, careful selection of sites may be necessary for obtaining correct estimates of HFD in fractured terrains. For example, in the present work the mean of seven HFD measurements within the study area is 65 ± 23 mW m^{-2} whereas the mean for five sites within the "cold zones" is 54 ± 13 mW m^{-2}, substantially lower than the mean of 95 ± 12 mW m^{-2}, for the two sites outside the "cold zones". Are such differences purely local phenomena or do they exist on a regional scale? An answer to this important question must await results of detailed HFD measurements accompanied by fracture studies in a large number of areas. On the other hand the difference between the two means, in spite of the obvious limitations of the data set, is a matter of concern. To judge from the pattern shown in figure (7), over fifty per cent of the area under investigation is found to be affected by interfracture fluid flows. It is not known at the moment whether such an estimate is representative of other hard rock terrains in continental regions.

It is tempting to speculate on possible implications of this observation on estimates of heat loss in continental areas. If the groundwater flows are weak and the perturbation effects are small, the distribution of HFD

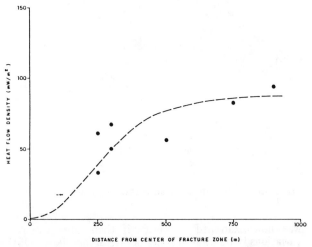

Fig. 8. Variation of heat flow with distance from the centre of the nearest "permeable" fracture.

sites in relation to the fracture zones is unimportant. In the case of moderate fluid flows, where the downgoing limbs of the flow path will be cooled and upper parts of upflow paths heated, a random distribution of HFD sites in relation to fracture zones may minimize systematic errors, provided data are obtained from depth intervals comparable to those of the heated upflow paths of circulation systems. Paradoxically, data from deeper levels may lead to underestimates of true HFD except, of course, in cases where data refer to depths below the circulation level. Conversely, if the fluid flow is substantial, the entire flow path is cooled and sites of HFD measurements have to be selected carefully to avoid systematic error. It is obvious that in regions of interfracture fluid flows the areal extent of "cold zones" is larger than that of "warm zones". There is hence a reasonably good probability that more low HFD values will be encountered than high ones, with the result that true HFD tends to be underestimated. The mean HFD values for continental regions reported in the literature (see, for example, Sclater et al, 1980) are based on data uncorrected for the effects of fluid flow in fractures. In this context we note that the mean HFD of 53 ± 13 mW m^{-2} for the five sites within the "cold zones" is in better agreement with the value given by Sclater et al (1980) for continental regions of Paleozoic to late Precambrian in age, than the overall mean of 65 ± 23 mW m^{-2} when sites outside the "cold zones" are included. In other words if such differences are indeed true, and not artifacts of limited data sets, there could be systematic errors of 10 to 20 per cent in the heat loss estimates for continental hard rock terrains.

Acknowledgements. The present study was carried out as part of an M.Sc. thesis work by one of the authors (A.C.D.). Financial support was provided by Fundacao Amparo a Pesquisa do Estado de Sao Paulo (FAPESP). Auxiliary facilities were provided by Instituto de Pesquisas Tecnologicas do Estado de Sao Paulo (IPT). We are thankful to our colleagues Dr. Alcides Frangipani and Dr. Fernando B. Ribeiro for fruitful discussions on the subject of interfracture fluid flows and Dr. Sundaram S.S. Iyer for critical comments on an earlier version of this manuscript.

References

Almeida, F.F.M., G. Amaral, U.G. Cordani, and K. Kawashita, The Precambrian evolution of the South American cratonic margin south of the Amazon river, in *The Ocean Basins and Margins*, edited by A.E.M. Nairn and F.G. Stehli, v.1, pp.411–446, 1973.

Almeida, F.F.M., Y. Hasui, Davino, A., and N.L.E. Haralyi, informacoes geofisicas sobre oeste mineiro e seu significado geotectonico, *Ann. Acad. Bras. Ci.*, 49–60, 1980.

Avery, T.E., *Interpretation of aerial photographs*, Burgess Publishing Co., Minneapolis, 1962.

Campnha, G.A.C., L.A. Fernandes, and A. Gimenes Filho, Quadriculas Mogi–Guacu e Aguas de Lindoia, *Atas da Primeira Jornada Sobre a Carta Geologica do Estado de Sao Paulo em 1:50,000*, 137–165, 1983.

Fiori, A.P., E. Wernick and J.S. Bettencourt, Evolucao policiclica na regiao nordeste do estado de Sao Paulo e

areas vizinhas do estado de Minas Gerais, in *30th Congresso Bras. Geologia*, v.1, 309–320, 1978.

Grossi Saad, J.H. and A.L.M. Barbosa, Sintese geologica de folha de Socorro, Sao Paulo, *Atas da Primeira Jornada Sobre a Carta Geologica do Estado de Sao Paulo em 1:50,000*, 109–134, 1983.

Hasui, Y., C.D.R. Carneiro and C.A. Bistrichi, Estruturas e tectonica do Precambriano de Sao Paulo e Parana. *Ann. Acad. Bras. Ci.*, 52, 61–76, 1980.

Hasui, Y., C.D.R. Carneiro and A.M. Coimbra, The Ribeira fold belt, *Rev. Bras. Geoc.*, 5, 257–266, 1975.

Hasui, Y., A.S.L. Dantas, C.D.R. Carneiro, and C.A. Bistrichi, O embasamento Precambriano e o Paleozoico em Sao Paulo, in *Mapa Geologico do Estado de Sao Paulo*, v.1, Instituto de Pesquisas Tecnologicas (IPT), 12–45, 1981.

Horai, K.I. and G. Simmons, Thermal conductivities of rock–forming minerals, *Earth Plan. Sci. Lett.*, 6, 359–368, 1969.

Hut, J.R. and J.W. Berg Jr., Thermal and electrical conductivities of sandstone rocks and ocean sediments, *Geophysics*, 38, 489–500, 1968.

Jones, F.W. and E.R. Oxburgh, Two dimensional thermal conductivity anomalies and vertical heat flow variations, in *Terrestrial Heat Flow in Europe*, edited by V. Cermak and L. Rybach, pp.98–106, Springer–Verlag, Berlin, 1979.

Lewis, T.J. and A.E. BEck, Analysis of heat flow data — detailed observations in many holes in a small area, *Tectonophysics*, 41, 41–59, 1977.

Long, J.C.S. and P.A. Witherspoon, The relatonship of the degree of interconnection to permeability in fracture networks, *J. Geophys. Res.*, 90,. 3087–3098, 1985.

Marangoni, Y.R., *Estudo comparativo entre metodos de medida de condutividade termica em materiais geologicos*, M. Sc. Thesis, Univ. of Sao Paulo, pp.174, 1986.

Marangoni, Y.R. and Del Rey, C.A., Condutividade termica de amostras do complexo Amparo, *Rev. Bras. Geofisica*, 4, 61–71, 1986.

Ricci, M. and S. Petri, Principios de Aerofotogrametria e interpretacao geologica, *Campanha Editora Nacional*, Sao Paulo, pp. 219, 1965.

Sagar, B. and A. Runchal, Permeability of fractured rock: Effect of fracture size and data uncertainties, *Water Resour. Res.*, v.18, 2, 266–274, 1977.

Sclater, J.G., C. Jaupart and D. Galson, The heat flow through oceanic and continental crust and the heat loss of the earth, *Rev. Geophys. Space Phys.*, 18, 264–311, 1980.

Snow, D.T., *A parallel plate model of fractured permeable media*, Ph.D. Thesis, Univ. of California, Berkeley, pp. 33, 1965.

Turcotte, D.L. and G. Schubert, *Geodynamics — Applications of continuum physics to geological problems*, pp. 450, Wiley, New York, 1982.

Witherspoon, P.A., Flow of groundwater in fractured rocks, *Bull. Int. Assoc. Eng. Geology*, 34, 103–115, 1986.

Woodside, W. and J.H. Messmer, Thermal conductivity of porous media. 1 — Unconsolidated sands, *J. Appl. Phys.*, 32, 1688–1699, 1961.

SUBSURFACE CONDUCTIVE AND HYDROGEOLOGICAL RELATED THERMAL EFFECTS IN THE AREA OF THE CRUSTAL ELECTRICAL CONDUCTORS OF THE WILLISTON BASIN

J. A. Majorowicz, F. W. Jones and M. E. Ertman

Institute of Earth and Planetary Physics and Department of Physics,
University of Alberta, Edmonton, Canada, T6G 2J1

Extended Abstract. Heat flow estimates for the Mesozoic sediments of the Williston Basin show a strong correlation with ground surface topography and hydraulic head and so the conductive heat flux from the crystalline crust and upper mantle is masked. The observed correlation has been interpreted as due to the effect of water motion which occurs in permeable Lower Mesozoic aquifers (Majorowicz et al. 1986; Gosnold, 1985). The lack of a similar correlation for the heat flow through deeper Paleozoic formations suggests less influence by the gravity driven hydrodynamic effect there and deep sedimentary heat flow anomalies correlate with deep crustal structures found by other independent methods. The elongated heat flow anomaly found in southern Saskatchewan from heat flow studies in Paleozoic strata (Majorowicz et al. 1986) extends into North Dakota as reported by Price et al. (1986) who based his observations on S_2 pyrolysis peak studies. The anomaly shown in Figure 1 lies east of the North American Central Plains electrical conductivity anomaly (NACP) as defined by Alabi et al. (1975). However, it coincides with electrical conductivity anomalies observed from magnetotelluric studies by Jones and Savage (1986) and Rankin and Pascal (pers. comm., 1987).

Anomalous structural and geophysical features in the basement and Phanerozoic strata are also present in the region where the heat flow and magnetotelluric anomalies occur (Majorowicz et al., 1988).

The source of the magnetotelluric and thermal features lies in the upper crust (Jones and Savage, 1986; Majorowicz et al., 1988). A numerical model which simulates the effect of enhanced heat generation in the upper crust

Copyright 1989 by
International Union of Geodesy and Geophysics
and American Geophysical Union.

related to mineralization or enhanced conductivity due to saline brine movement along interconnected fractures is proposed. The geometry of the high electrical conductivity structure is taken from Rankin and Pascal,

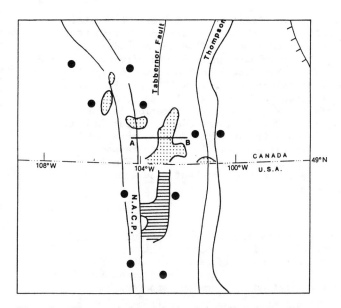

Fig. 1. The position of the heat flow anomaly in the Williston Basin as defined by Majorowicz et al. (1986) from quantitative heat flow estimates in Paleozoic sediments (Q>100mWm) in Canada and quantitative heat flow anomaly location estimates from S_2 Pyrolysis peak values (according to Price, et al., 1986) in the U.S. The location of the North American Central Plains electrical conductive body as mapped from magnetometer work is shown (NACP). Locations of Alabi et al. (1975) array magnetometers are indicated by dots. Major tectonic lineaments are shown (according to Green et al., 1985).

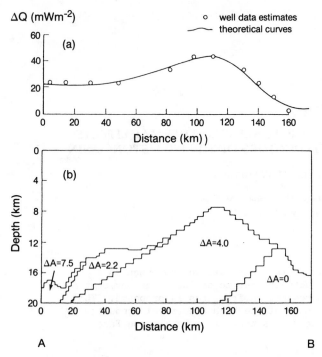

Fig. 2. Two dimensional model of the crustal structure and the numerically calculated heat flow effects along profile A-B as indicated in Fig. 1. Assumed: uniform conductivity $\lambda = 2.5$ W m^{-1} K^{-1}, heat flow into the bottom of the model $Q_m = 0$, density $\sigma = 2.76 \times 10^3$ kg m^{-3}, specific heat capacity $C = 1.3 \times 10^3$ J kg^{-1}, ΔA is the difference between heat generation in the particular block of the deep structure and the crust above it (in μW m^{-3}). The observed heat flows (dots) are normalized with respect to an assumed regional heat flow of 60 mW m^{-2}. Average uncertainty of heat flow estimates is 25 percent.

(ibid), and is similar to the anticlinal high electrical conductive body in the upper crust as proposed by Jones and Savage, (1986). The anticlinal structure has been assigned anomalous heat generation or anomalous apparent heat conductivity. The fit of the predicted heat flow anomaly with the heat flow observations from wells is poor. However, a close fit of the observations can be achieved when the structure

is divided into blocks characterized by different anomalous heat generation. The model predicts high heat generation in the mid-upper crust along profile A-B. The preferred possible cause for the anomalous heat flow is enhanced heat generation in the upper crust (see Fig. 2), possibly related to mineralization.

Interpretation of the heat flow anomaly in relation to the heating and thinning of the lithosphere as a result of its rejuvenation when the basin was initiated is uncertain at this stage. That event may have caused redistribution of radiogenic elements and high electrical conductivity mineralization. However, the proposed crustal high heat generation zone may result from redistribution of radiogenic elements during the tectonic development of the Superior-Churchill boundary zone. In this latter case, the zone of weakness in the lithosphere may have existed with its location approximately coincidental with the center of the later formed Williston basin.

References

Alabi, A. O., Camfield, P. A. and Gough, D. I., The North American Central Plains conductivity anomaly, Geophys. J. R. Astr. Soc., 43, 815-835, 1975.
Gosnold, W. D., Heat flow and ground water flow in the great plains of the United States, J. Geodynamics, 4, 247-264, 1985.
Green, A. G., Hajnal, Z. and Weber, W., An evolutionary model of the Western Churchill Province and western margin of the Superior Province in Canada and the north-central United States, Tectonophysics, 116, 281-322, 1985.
Jones, A. G., and Savage, P. J., North American Central Plains conductivity anomaly goes east, Geophys. Res. Letters, 13, 685-688, 1986.
Majorowicz, J. A., Jones, F. W. and Jessop, A. M., Geothermics of the Williston Basin and hydrocarbon occurrence, Geophysics, 51, 767-779, 1986.
Majorowicz, J. A., Jones, F. W. and Osadetz, K. G., Heat flow environment of the electrical conductivity anomalies in the Williston basin and hydrocarbon occurrence, Bull. Can. Petr. Geol., 36, 86-90, 1988.
Price, L. C., Davies, T., and Pawlewicz, M., Organic metamorphism in the lower Mississippian - Upper Devonian Bakken shales, Part 1: Rock-evalpyrolysis and vitrinite reflectance, J. Petr. Geol., 9, 125-162, 1986.

MAPPING THERMALLY DRIVEN UPFLOWS BY THE SELF-POTENTIAL METHOD

T. Ishido, T. Kikuchi and M. Sugihara

Geological Survey of Japan, Higashi 1-1-3, Tsukuba, 305 Japan

Abstract. Self-potential (SP) surveys have
been carried out of a number of geothermal areas
in Japan during the last decade. In most cases,
SP anomalies of positive polarity (50 to 500 mV
in amplitude and 0.5 to 5 km in spatial extent)
were found to overlie high temperature upflow
zones. The streaming potential generated by
hydrothermal circulation is believed to be the
most likely cause of the observed positive
anomalies. Whether or not subsurface upflows
produce observable potentials at the surface
depends on (among other things) the temperature,
the chemical composition of the pore water, and
the upflow rate. In addition to SP anomalies under natural state conditions, repeated surveys
of the Nigorikawa caldera area in Japan detected
a change in SP induced by production of geothermal fluids. The observed change is dipolar and
can also be attributed to an electrokinetic
mechanism.

Introduction

Self-potential (SP) anomalies of widely varying amplitude, polarity, and spatial extent have
been reported from several geothermal areas
[Zohdy et al., 1973; Zablocki, 1976; Anderson and
Johnson, 1976; Corwin and Hoover, 1979]. Although the various mechanisms involved in
producing these surface potentials have not been
positively identified, an electrokinetic process
related to the upward movement of pore fluids is
believed to be the cause of the anomalies of
positive polarity observed in Hawaii [Zablocki,
1976], Yellowstone [Zohdy et al., 1973], and Long
Valley [Anderson and Johnson, 1976].

The results of quantitative modeling of a
potential generating mechanism by hydrothermal
circulation were reported by Ishido [1981]. His
modeling is based on realistic values of the
electrokinetic coupling coefficients for crustal
rock-water systems estimated from experimental
results [Ishido and Mizutani, 1981; Ishido et
al., 1983]. In this paper, Ishido's [1981] work

Copyright 1989 by
International Union of Geodesy and Geophysics
and American Geophysical Union.

will be explained briefly, and examples from
recent self-potential surveys carried out in
Japan will be provided. Next, after the discussion of the SP anomalies under natural state conditions, a change in SP induced by the production
of geothermal fluids will be described.

Streaming Potential Associated with Hydrothermal Circulation

The flow of a fluid through a porous medium
may generate an electrical potential gradient
(called the electrokinetic or streaming
potential) along the flow path by the interaction
of the moving pore fluid with the electrical
double layer at the pore surface, a process known
as electrokinetic coupling. The phenomenological
equations describing electrokinetic coupling in a
porous medium are given by Ishido and Mizutani
[1981]:

$$I = -\eta t^{-2}\sigma\nabla\phi + \eta t^{-2}(\epsilon\zeta/\mu)(\nabla p - \rho g) \quad (1)$$

$$J = \eta t^{-2}(\epsilon\zeta/\mu)\nabla\phi - (k/\mu)(\nabla p - \rho g) \quad (2)$$

where I is the electric current density, J the
fluid volume flow flux, ϕ the electric potential,
p the pore pressure and g the acceleration due
to gravity. ρ, ϵ, and μ are the density,
dielectric constant, and dynamic viscosity of the
pore fluid, respectively; η, t, and k are the
porosity, tortuosity, and permeability of the
porous medium, respectively. $\sigma = \sigma_f + m^{-1}\sigma_s$ (σ_f
is the pore fluid electrical conductivity and σ_s
the surface electrical conductivity; and m,
hydraulic radius) and ζ is the "zeta potential",
the potential across the electrical double layer.
Substituting (2) into (1) yields :

$$I = -\eta t^{-2}\sigma[\{1-\eta t^{-2}(\epsilon^2\zeta^2/\sigma\mu k)\}\nabla\phi + C(\mu/k)J] \quad (3)$$

where $C(= \epsilon\zeta/\sigma\mu)$ is called the streaming potential coefficient. If C is negative (positive),
the positive (negative) charge is carried by the
fluid flow J. For geologic materials the quan-
tity $\eta t^{-2}(\epsilon^2\zeta^2/\sigma\mu k)$ is safely neglected ($O(10^{-5})$
in the following case). In the absence of cur-

rent sources, $\nabla \cdot I = 0$; and for homogeneous regions, using (3):

$$\nabla^2 \phi = -C(\mu/k)\nabla \cdot J \qquad (4)$$

Considering thermally driven convection in single-phase (liquid) systems, $\nabla \cdot J = 0$ (Boussinesq approximation), we thereby obtain,

$$\nabla^2 \phi = 0 \qquad (5)$$

Sources for ϕ therefore can only occur at boundaries. Continuity of normal current flow requires

$$n \cdot I_1 = n \cdot I_2 \qquad (6)$$

where n is the unit vector normal to the boundary.

When a hydrothermal convection cell is confined underground so that no fluid flow intersects the ground surface, no surface electric potential anomaly will appear for a uniform half space, since the solution of (5) without any sources for ϕ must be zero throughout the half space. This is not the case, however, for inhomogeneous media where there are boundaries between regions of differing physical properties such as C and electric charge can be accumulated by fluid flow at the boundaries.

The streaming potential coefficient (C) is presumably distributed heterogeneously in the earth. Experimental results [Ishido and Mizutani, 1981; Ishido et al., 1983] show that the principal parameters affecting the zeta potential (ζ) and/or the streaming potential coefficient in silicate rock-water systems are the pH of the aqueous solution, the electrolyte concentration of the solution and the temperature of the system. The ζ potential (and hence C) will be negative if the pH is higher than about 2, and will increase in magnitude with decreasing electrolyte concentration and/or increasing temperature. When a small amount (about 1 ppm) of hydrolyzable metal ions such as Al^{3+} or Fe^{3+} are present in the solution, the ζ potential (and C) becomes positive at temperatures below about 100°C. The experimental results described above imply that C will be inhomogeneous in a hydrothermal convection cell in which there is a large temperature contrast.

Using the values of C estimated from the experiments [Ishido and Mizutani, 1981; Ishido et al., 1983], Ishido [1981] developed quantitative models of the electric potentials generated by hydrothermal circulation through electrokinetic coupling. Equation (5) was solved numerically subject to boundary condition (6). In one of the models (Figure 1), a half space below the surface was divided into two regions: one characterized by temperatures above 150°C and the other by lower temperatures. Physical properties of the

higher and lower temperature regions were assumed to be those at 200°C and 100°C, respectively. Considering a typical crustal rock-water system containing water with pH = 7, 0.02 mol 1^{-1} NaCl, and 10^{-5} mol 1^{-1} Al^{3+}, Ishido [1981] estimated the appropriate values of C as -35 and 0 mV bar^{-1} for the higher and lower temperature regions, respectively (see also Ishido and Mizutani [1981]). As shown in Figure 1, positive electric potential appears around the portion of the thermal boundary intersected by outward fluid flow from the higher temperature region. The accumulation of positive charge at this por-

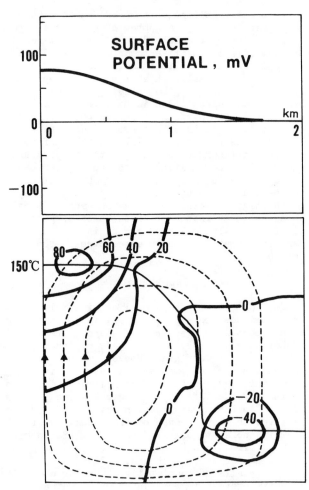

Fig. 1. The lower diagram shows the electric potential distribution (solid lines, in mV) generated by hydrothermal circulation (stream lines are shown by broken lines). The entire region is 2x2 km^2 and divided into two subregions by the 150°C isotherm. Physical properties of the higher and lower temperature regions were assumed to be those at 200°C and 100 °C, respectively. The upper diagram shows the SP distribution on the earth's surface. (After Ishido, 1981)

tion of the boundary is caused by fluid flow carrying positive charge (C < 0) and no charge (C = 0) in the higher and the lower temperature regions, respectively. The opposite effect produces an accumulation of negative charge in the lower right corner of the diagram.

Ishido [1981] has shown that an observable self-potential anomaly (10-100 mV in magnitude) can appear at the surface if the following conditions are satisfied: (1) the circulating fluid is an aqueous solution of neutral pH(> 4) and moderate concentration of dissolved salt (< 0.1 mol l^{-1}), and (2) the fluid volume flow flux (J) is larger than $10^{-8} - 10^{-7}$ m s^{-1}. The polarity of the anomaly over a hot zone is always positive whether or not the fluid flow (with nonzero C) intersects the surface ; this is mainly because C is negative (ζ < 0) and larger in magnitude under high temperature conditions.

Results of Recent SP Surveys in Japan

During the last decade self-potential (SP) surveys were conducted in six Japanese geothermal areas by the Geological Survey of Japan (GSJ) and in four additional areas by the New Energy Development Organization (NEDO). Each survey covers an area of 50-100 km^2 with survey lines of about 100 km in total length. SP anomalies of various types have been recorded through these surveys, and obvious anomalies of positive polarity were found in seven different areas: the Kutcharo and Nigorikawa calderas in Hokkaido island, the Sengan and Okuaizu geothermal areas in the northern part of Honshu island, and the Hohi, Unzen and Kirishima geothermal areas in Kyushu island. In most cases, the correlation between the anomaly of positive polarity and the high temperature upflow zone is evident. Two examples are described here.

Mt. Yake Volcano Area

Mt. Yake (Yake-yama), located in the Sengan geothermal area, northern Honshu, is an active volcano and has many thermal manifestations such as fumaroles, hot springs and alteration zones. Under a joint program, GSJ and NEDO have been conducting geological, geochemical, and geophysical surveys and also conducting an exploratory drilling program in the Sengan geothermal area. SP surveys in the Mt. Yake area were carried out by GSJ starting in 1979. Exploration efforts were also undertaken by a private developer in the area; Mitsubishi Metal Corporation (MMC) has been operating a 10 MW geothermal power plant at Ohnuma since 1974 and conducting an exploratory well drilling program in the Sumikawa field since 1981.

Surface rocks in the area shown in Figure 2 are mostly andesitic lavas from Mt. Yake, 1336 m in elevation above sea level. The elevations of the fumaroles and hot springs shown in Figure 2

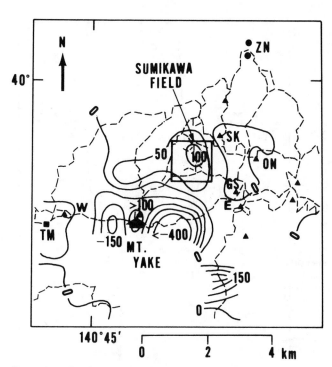

Fig. 2. Self-potential distribution in the Mt. Yake area. Contour interval is 50 mV. Survey lines are shown by broken lines. Areas with fumaroles and/or steam-heated sulfate type springs are shown by triangles (SK, Sumikawa; ON, Ohnuma; GS, Goshogake). Areas with chloride-bicarbonate type thermal water discharge are shown by solid circles (ZN, Zenikawa). Tamagawa (TM) hot springs area is shown by a solid square.

are generally from 700 to 1100 m except for the fumaroles near the volcano summit and for the Zenikawa bicarbonate hot springs to the north (520 m). The most remarkable feature of the thermal waters of the area is their compositional variety. Waters discharged from production wells in the Ohnuma field and from exploratory wells in the Sumikawa field have neutral pH and total dissolved salt contents of about 2000 ppm on the average [Sakai and Mori, 1981].

Self-potential measurements were made with copper-copper sulfate nonpolarizing electrodes and a high-impedance voltmeter. For each survey line, the maximum wire length (from a fixed base electrode) and the data sampling intervals were 2000 and 100 meters respectively. Telluric activity was monitored by recording potentials across stationary dipoles in the survey area ; no significant telluric variation was observed during the survey period (in November, 1983). Figure 3 shows an example of profile data obtained along the survey line E-W which traverses the summit of Mt. Yake (Figure 2). Also shown is the data obtained two years later (in October,

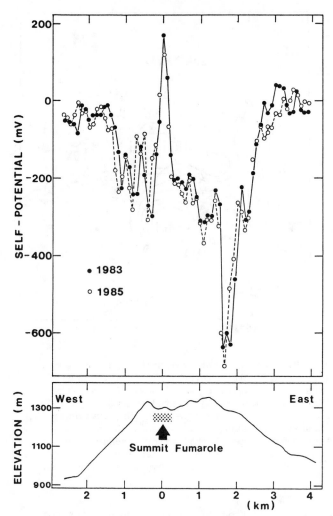

Fig. 3. Self-potential profiles, line E-W, Mt.
Yake area, measured two years apart. ●- 1983
data, ○- 1985 data. (The location of line E-W is
shown in Fig. 2.)

volcano summit. The positive anomaly to the
north of Mt. Yake overlies the Sumikawa field,
where the subsurface temperature is higher than
the surroundings (over 250°C at sea level) and a
vapor dominated zone is found at depths of 400-
600 m above sea level [Kubota, 1985]. Considering the chemistry of the subsurface waters in the
Sumikawa field, we believe that the subsurface
upflow carries enough electric charge (|C| is
large enough) to produce the observed SP anomaly
through electrokinetic coupling. As the temperature of the upflow decreases from 300°C (maximum
temperature observed in the field) to below 200°C
along the flow path, the streaming potential
coefficient (C) will probably decrease in
magnitude; therefore, positive charge (C<0)
carried by the upflow will be left where the substantial decrease in temperature occurs. The
model presented in the previous section (Figure
1) appears to explain the generation of self-
potential in the Sumikawa field.

The anomaly observed at the volcano summit,
over 200 mV in magnitude and 500 m in width
(Figure 3), is thought to be generated by a subsurface upflow which causes the fumarolic activity at the surface. Since electric charge
cannot be carried by the flow of the steam phase
alone [Marsden and Tyran, 1986], positive charge
carried by the hot water upflow from depth will
probably accumulate around a zone where separation between the vapor and the liquid phases occurs. The chemistry of the hot waters is expected to yield a large enough C value to explain the observed anomaly. We do not have sufficient exploration data to discuss correlations
between the anomaly and geothermal activity in
the area to the southeast of Mt. Yake.

As seen in Figure 2, there is no obvious
anomaly overlying the hot spring areas. This is
mainly because the thermal waters in these areas
are characterized by very small C values under
near-surface conditions and generate no
measurable anomaly associated with their discharge at the surface. The thermal waters from
areas such as Tamagawa (TM), Sumikawa (SK) or
Goshogake (GS) (Figure 2) have low pH (= 1 - 3) ;
the streaming potential coefficients (C) for
those waters are expected to be very low at temperatures below 100°C [Ishido and Mizutani,
1981]. As described above, the production of
geothermal fluids has been undertaken at Ohnuma,
indicated by "ON" in Figure 2. Production-
induced SP changes will be discussed later.

There are two obvious anomalies of negative
polarity on the eastern and western flanks of Mt.
Yake. The downflow of meteoric waters can generate such negative anomalies through
electrokinetic coupling [Ishido,1981]. If the
flow rate (J) of the downflow is 10^{-7} m s^{-1}, a
negative anomaly over 100 mV in magnitude will
appear on the surface. We do not, however, have
enough measured data to warrant further discussion of these features.

1985), by open circles. As may be seen in
Figure 3, data reproducibility is fairly good,
particularly for SP anomalies greater than 500 m
in spatial extent.

In compiling the contour map (Figure 2) showing the distribution of the self-potential in Mt.
Yake area, the potentials measured along the
various survey lines were tied to a common ground
reference and then subjected to a smoothing
process (variations shorter than about 500 m in
wavelength were removed). Closure offsets were
relatively small: for example, a 48.7 km loop
traverse was closed with an error of 5.0 mV.

As shown in Figure 2, anomalies of positive
polarity appear in three areas: on the northern
and southeastern flanks of Mt. Yake, and at the

Okuaizu Geothermal Field

The Okuaizu geothermal field is located in Fukushima prefecture on the Japanese island of Honshu. During 1982-84 NEDO completed geological, geochemical, and geophysical explorations including an SP survey over an area of about 70 km^2, and drilled seven 1000-1500 m depth wells and six about 400 m depth heat flow holes. After NEDO's survey, Okuaizu Geothermal Co., Ltd. (OAG) conducted an exploratory well drilling program in the area.

The area of investigation has a mean elevation of approximately 300-500m above sea level and is dominated by 729 m Mt. Yuno (Yuno-take). The temperature distribution at -500 m ASL (that is, 500 m below sea level) is shown in Figure 4. The high temperature zone is located around wells OA-4 and 6; the highest temperature observed in the field is 286°C at -1120 m ASL in well OA-6. Well OA-4 was discharged for a month in 1984 and pressure transient tests were carried out [Ishido, 1985].

Also shown in Figure 4 is the result of the SP survey; a positive anomaly of about 100 mV overlies the shaded area shown in Figure 4. It

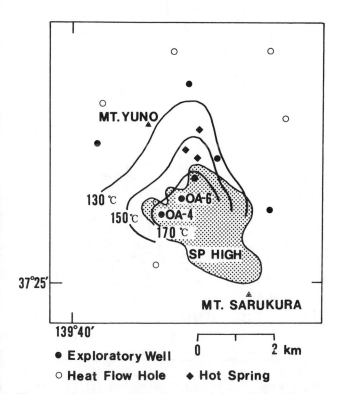

Fig. 4. Geothermal features and locations of wells in the Okuaizu field. Contours show temperature distribution at -500 m above sea level. SP anomaly of positive polarity (about 100 mV) overlies the shaded area.

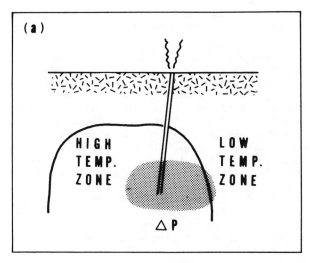

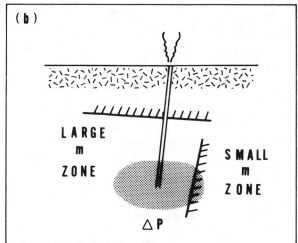

Fig. 5. Models for production-induced SP change. (a) High-low temperature interface, or (b) large-small hydraulic radius (m) interface acts as a boundary between regions of differing streaming potential coefficient. Zone of pressure change (Δp) induced by production is shown by shade.

should be noted that the northern part of the area of positive anomaly coincides with the locations of high temperature and high productivity wells drilled by OAG (locations of OAG's wells are not shown in Figure 4). Waters discharged from the exploratory wells have neutral pH and total dissolved salt contents of about 20,000 ppm [Nitta et al., 1987]. The streaming potential coefficient for those waters is expected to be small (about 1 mV bar^{-1} in order of magnitude) even under high temperature conditions because of high salinity (about 0.3 mol l^{-1} NaCl) [Ishido and Mizutani, 1981; Ishido et al., 1983]. Volume fluxes (J) exceeding 10^{-7} - 10^{-6} m s^{-1} are probably required for upflows to produce the observed anomaly through electrokinetic coupling.

SP Change Induced by Production of Geothermal Fluids

When a sink or source of fluid is present within a reservoir as a result of production of geothermal fluids, a surface electric potential anomaly can be produced through electrokinetic coupling if the following conditions (which have been derived for electrokinetic effects represented by equations (1) and (2) without buoyancy terms) are satisfied. First, there must be a boundary separating regions of differing streaming potential coefficient C; second, there must be a nonzero component of pressure gradient parallel to this boundary [Nourbehecht, 1963; Fitterman, 1978].

A temperature boundary, a boundary between regions of different pore water chemistry, and/or a contact of different rock formations represent the most probable candidates for discontinuities in the value of C in a geothermal reservoir (Figure 5). Since the value of C is sensitive to temperature, the interface between the high temperature region within the reservoir and the cooler surroundings acts as a boundary between regions of differing C (Figure 5(a)). When two kinds of rocks come into contact with each other and other pertinent parameters such as pore water chemistry are homogenous, no major deviations in C are expected for crustal rocks [Ishido and Mizutani, 1981]. However, if the value of hydraulic radius (m) of the pores and the cracks varies significantly across the boundary, a large difference in C can appear [Ishido and Mizutani, 1981]. A discontinuity in pore hydraulic radius can therefore provide a boundary between regions of differing C (Figure 5(b)).

When a pressure change induced by production and/or injection of fluids propagates to a boundary between regions of differing C in a reservoir (Figure 5), an electric potential anomaly appears at the earth's surface through electrokinetic coupling. We can model the potential generating mechanism quantitatively using the formulation derived by Nourbehecht [1963] and Fitterman [1978,1979]. If the boundary is nearly vertical, the anomaly will be dipolar in waveform [Fitterman,1979].

Production-induced Electric Potential Observed at Nigorikawa

The Nigorikawa caldera is located in the southern part of Hokkaido island, Japan. The diameter of the caldera is about 3 km and various fumaroles and hot springs are located in the northern half of the caldera floor. The thermal waters are neutral and contain substantial HCO_3^- (500-1000 ppm). The concentration of NaCl ranges from 10^{-3} to 10^{-1} mol l^{-1}. The Mori geothermal power plant (50 MW) was built in 1982 and has been in continuous operation since. SP surveys were conducted by GSJ three times: in 1978 and 1981 (before plant startup) and in 1984 after the operation of the power plant began.

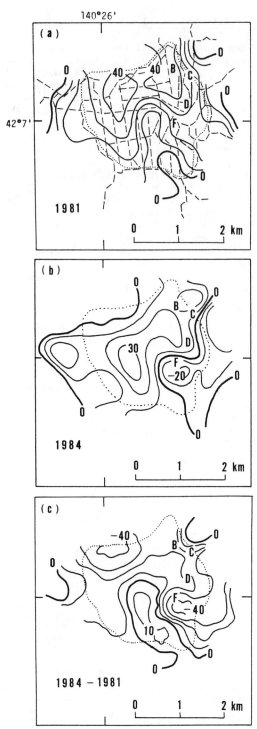

Fig. 6. Self-potential distributions in the Nigorikawa caldera, Japan observed (a) in 1981, and (b) in 1984. (c) Difference in SP distribution between 1981 and 1984. Contour interval is 10 mV. Broken lines shown in Fig.(a) denote survey lines used for 1981 and 1984 surveys (the data sampling intervals are 100 m). The edge of the caldera floor is shown by dotted lines. The well sites are shown as B - F.

The result of the 1981 SP survey is shown in Figure 6(a); SP is high inside the caldera compared to the surrounding area and anomalies of positive polarity overlie the northern area of hot springs. The results obtained from the 1978 survey show virtually the same potential distribution as the 1981 survey in the caldera; the anomalies greater than about 500 m in spatial extent are similar for the two surveys. The positive anomalies are thought to be generated by natural upflows through electrokinetic coupling [Ishido,1981]. However, as can be seen from Figure 6, the SP distribution observed in 1984 is significantly different from that in 1981 (and that in 1978). The difference between the 1981 and 1984 SP distributions is shown in Figure 6(c). The dipolar change in SP, which is comparable in magnitude to the "natural" SP observed in 1981 (Figure 6(a)), appears over the principal zone of fluid production. This observed change is believed to be generated by underground fluid flows resulting from the production (and reinjection) of geothermal fluids through electrokinetic coupling. Quantitative modeling of the induced SP change has been carried out by Kikuchi et al. [1986] on the basis of a formulation similar to that of Fitterman [1979].

We have also carried out SP surveys in several other areas containing operating geothermal power plants in addition to Nigorikawa (22 MW), Kakkonda (50 MW) and Ohnuma (10 MW; see Figure 2) in the Sengan geothermal area; Onikobe (12 MW) in the Kurikoma thermal area; and Ohtake (12 MW) and Hatchobaru (55 MW) in the Hohi thermal area. Since the possibility always exists that production-induced SP changes overlie the natural self-potential distribution associated with the undisturbed state, care must be exercised in interpreting SP data from areas such as those mentioned above.

Conclusions

The correlation between positive SP anomalies and high temperature upflow zones has been confirmed for several geothermal areas in Japan. Ishido's [1981] quantitative models of the electrokinetic effect show that the streaming potential generated by hydrothermal circulation is the most likely cause of the observed positive anomalies. The anomaly which results from natural hydrothermal circulation may or may not be observable depending upon the temperature, chemical composition of the pore water, flow rate, and geometry of the hydrothermal convection cell.

In the neighborhood of an operating geothermal power plant, there is the possibility of production-induced SP changes through electrokinetic coupling. In order to delineate SP anomalies generated by natural thermally driven upflows, we must consider the effect of production-induced SP changes in addition to other effects such as nonthermal subsurface

water flow, conductive mineral deposits, and soil chemistry variations [Corwin and Hoover,1979]. The self-potential method appears to be a promising technique for the detection and mapping of thermally driven upflows in geothermal areas.

Acknowledgments. This work was supported by the MITI's "Sunshine" project under the direction of the Geological Survey of Japan, Geothermal Research Department. We thank K. Ogawa and K. Kimbara for their constant encouragement. K. Baba is gratefully acknowledged for his guidance during the early stages of this work.

References

Anderson, L.A., and G.R.Johnson, Application of the self-potential method to geothermal exploration in Long Valley, California, J. Geophys. Res., 81, 1527-1532, 1976.

Corwin, R.F., and D.B.Hoover, The self-potential method in geothermal exploration, Geophysics, 44, 226-245, 1979.

Fitterman, D.V., Electrokinetic and magnetic anomalies associated with dilatant regions in a layered earth, J. Geophys. Res., 83, 5923-5928, 1978.

Fitterman, D.V., Calculations of self-potential anomalies near vertical contacts, Geophysics, 44, 195-205, 1979.

Ishido, T., Streaming potential associated with hydrothermal convection in the crust: a possible mechanism of self-potential anomalies in geothermal areas (in Jap., with Engl. abstr.), J. Geotherm. Res. Soc. Jpn., 3, 87-100, 1981.

Ishido, T., Pressure transient tests at the Okuaizu geothermal field in Japan, Geothermal Resources Council, Transactions, Vol.9-Part II, 521-526, 1985.

Ishido, T., and H. Mizutani, Experimental and theoretical basis of electrokinetic phenomena in rock-water systems and its applications to geophysics, J. Geophys. Res., 86, 1763-1775, 1981.

Ishido, T., H. Mizutani, and K. Baba, Streaming potential observations, using geothermal wells and in situ electrokinetic coupling coefficients under high temperature, Tectonophysics, 91, 89-104, 1983.

Kikuchi, T., M. Sugihara, and T. Ishido, Self-potential change induced by geothermal fluid production (in Japanese), Abstr. Geotherm. Res. Soc. Jpn., 61, 1986.

Kubota, Y., Conceptual model of the north Hachimantai-Yakeyama geothermal area(in Jap., with Engl. abstr.), J. Geotherm. Res. Soc. Jpn., 7, 231-245, 1985.

Marsden, Jr., S.S., and C.K. Tyran, The streaming potential generated by flow of wet steam in capillary tubes, Proceedings of the 11th Workshop on Geothermal Reservoir Engineering, Stanford Univ., 1986.

Nitta, T., S.Suga, S. Tsukagoshi, and M. Adachi, Geothermal resources in the Okuaizu, Tohoku district, Japan (in Jap., with Engl. abstr.), J. Jpn. Geotherm. Energy Assoc., 24, 340-370, 1987.

Nourbehecht, B., Irreversible thermodynamic effects in inhomogeneous media and their applications in certain geoelectric problems, Ph.D. thesis, M.I.T., 1963.

Sakai, S., and H. Mori, Ohnuma geothermal field, in Field Excursion Guide to Geothermal Fields of Tohoku and Kyushu, Volcanological Soc. Jpn., pp.21-29, 1981.

Zablocki, C.J., Mapping thermal anomalies on an active volcano by the self-potential method, Kilauea, Hawaii, in Proc. 2nd U.N. sympos. on the Devel. and Use of Geothermal Resources, Vol.2, pp.1299-1309, 1976.

Zohdy, A.A.R., L.A. Anderson, and L.J.P. Muffler, Resistivity, self-potential, and induced-polarization surveys of a vapor-dominated geothermal system, Geophysics, 38, 1130-1144, 1973.